Electromagnetic Waves and Optics

Navin Khaneja

Electromagnetic Waves and Optics

A Linear Systems Approach

Springer

Navin Khaneja
System and Control Engineering
Indian Institute of Technology Bombay
Mumbai, Maharashtra, India

ISBN 978-3-031-76143-0 ISBN 978-3-031-76144-7 (eBook)
https://doi.org/10.1007/978-3-031-76144-7

This Springer imprint is published by the registered company Springer Nature Switzerland AG
The registered company address is: Gewerbestrasse 11, 6330 Cham, Switzerland

Dedicated to Darshan Lal Khaneja

Preface

The origin of this book can be traced back to author's research in nuclear magnetic resonance (NMR) spectroscopy. In 1945, Purcell, Torrey, and Pound detected Nuclear Magnetic Resonance (NMR) in condensed matter. They had a tank (cavity/resonator) they filled with radio-frequency (rf) waves and immersed nuclear spins in it, all placed in static magnetic field B. When B was tuned, spins absorbed rf, and water level in the tank (rf level in the cavity actually) went down. How do you fill tanks with rf/microwaves. What will matter is we fill it at right frequency, not the shape of the cavity or filling radiator. How do we understand all this. Modern NMR/radio equipment uses tank circuits to amplify weak signals and LC oscillators to generate rf. Radars use special oscillator/cavities to generate microwaves and special pipes called waveguides to channel them. In this book, we study all this in the framework of linear systems.

Maxwell equations that govern the evolution of EM waves is a linear system. To solve for the linear system, we have to calculate its eigenvaues and eigenfunctions (modes). In this book, we take this approach and calculate modes for Maxwell equations, both in free space (where they are traveling waves) to space bounded by conductors like cavities and waveguides (where they are standing waves). Central to the theme of the book is the phenomenon of *resonance*. When the modes of the linear system have imaginary eigenvalues $j\omega_0$, and we drive the system with a harmonic input $\exp(j\omega_0 t)$, the mode builds up unbounded. We say, we drive the system on resonance and pump energy into it. In this book, we show how we study cavities, resonators, and waveguides as systems driven on resonance. We study tank circuits and resonant antennas. We show resonance is central to how we produce so called microwaves, in cavities. Another important notion in the study of linear systems is the concept of a steady state. When eigenvalues of linear system have negative real parts then system is termed lossy and in response to a harmonic input settles down to a steady state satisfying an algebraic equation. This steady-state analysis forms the basis of analysis of antenna's and transmission lines as discussed in the text.

In this book, we study analysis of Maxwell equations, waveguides, cavities, transmission lines, and antennas. We study oscillators for producing rf/microwaves and amplifiers and tank circuits for amplifying these signals. The latter part of this book is devoted to the study of light including optical instruments and phenomenon of optical resonance in an atom, which explains dielectric response of a medium and absorption of light. In writing the book, I have tried to answer all the questions I had as an electrical engineering undergraduate at IIT Kanpur, taking an electromagnetism course. I have chosen subjects that I feel are conceptual. I have stressed system theoretic ideas. In using analysis tools of electromagnetic theory that can be found elsewhere, I have stressed what system theoretic concept we are invoking. May that be calculation of modes, or steady state of a linear system, notion of resonance and resonant driving, etc. The book is written as a research monograph, but it is more of an effort on author's part to see all of electromagnetism as a linear system. The book is intended for a first course in electromagnetism or electromagnetic theory. The book may also be used in a course on rf/microwave engineering and optics.

It is an opportunity to acknowledge numerous people who I believe directly and indirectly helped with this effort. I would like to thank Professor Steffen Glaser and Niels Nielsen for numerous years of excellent collaboration in NMR spectroscopy that ultimately led me to develop this text. I would like to thank Professor Roger Brockett, who helped me in nurturing my taste in physics. I am grateful to Professor M. Sachidananda at IIT Kanpur whose course on electromagnetism has kept my interest in the subject alive and culminated in this text. I will like to thank the wonderful colleagues and academic environment of SYSCON at IIT Bombay that provided ample opportunity for self-development. Finally I like to acknowledge the support of my family, which made this effort possible.

Mumbai, India
June, 2019

Navin Khaneja

Contents

Part I
Electromagnetic Waves and Linear Systems

Electromagnetic (EM) waves are all around us: from the world of light and color, to TV and radio broadcast, to cell phones and dish TV, everywhere. In this book, we study generation and propagation of EM waves as a linear system. The evolution of EM waves is governed by Maxwell equations. We treat Maxwell equations as a linear system, driven by current sources so called antennas. So what is a linear system. To fix ideas, consider a spring mass system of mass m and spring constant k, to which force F is applied. The displacement of the system from its equilibrium denoted by x satisfies

$$m\ddot{x} + kx = F, \tag{1}$$

or

$$\ddot{x} + \omega_0^2 x = F/m, \tag{2}$$

with $x_1 = \omega_0 x$ and $x_2 = \dot{x}$, and $u = \frac{F}{m}$ we find the above system written as

$$\frac{d}{dt}\begin{bmatrix} x_1 \\ x_2 \end{bmatrix} = \underbrace{\begin{bmatrix} 0 & \omega_0 \\ -\omega_0 & 0 \end{bmatrix}}_{A}\begin{bmatrix} x_1 \\ x_2 \end{bmatrix} + \underbrace{\begin{bmatrix} 0 \\ 1 \end{bmatrix}}_{b} u. \tag{3}$$

If we denote $X = \begin{bmatrix} x_1 \\ x_2 \end{bmatrix}$, we can write the above system as

$$\frac{dX}{dt} = AX + bu. \tag{4}$$

In general, A is an $n \times n$ matrix, X, b are $n \times 1$ vector, and u is a time varying input. We call above equation a *linear system*. Maxwell equations are a linear system, where inputs are current sources that act as antenna. To solve for evolution of linear system, we have to solve for the eigenvalues and eigenvectors of the matrix A. The eigenvectors are called *modes*. This is what we do in this book, we solve for modes of the Maxwell equations which in free space are traveling waves. We solve for modes of the Maxwell equations when the space is bounded by conductors as in what are called waveguides, cavities, and resonators. These are called cavity modes. Central to theme of the book is the phenomenon of *resonance*. When the modes of the linear system have imaginary eigenvalues $j\omega_0$ and we drive the system with a harmonic input $u = \exp(j\omega_0 t)$ or $u = \cos\omega_0 t$, the mode builds up unbounded. We say we drive the system on resonance and pump energy into it. In this book, we show how we study cavities, resonators, and waveguides as systems driven on resonance. We study tank circuits and resonant antennas. We show resonance is central to how we produce so-called microwaves in cavities. Another important notion in study of linear systems is the concept of a steady state. When eigenvalues of A have negative real parts then system is termed lossy and in response to an input $u = \exp(j\omega_0 t)$ settles down to a steady state vector $X_0 \exp(j\omega_0 t)$ with X_0 satisfying the linear equation,

$$j\omega_0 X_0 = aX_0 + b \tag{5}$$

This steady-state analysis forms the basis of analysis of antennas and transmission lines as discussed in the text. Even when system is lossless, there is always a small resistance present making it lossy and we can talk about steady state. While the earlier part of the book covers EM waves as low frequencies radio waves (3 kHz–3 GHz) and microwaves (3 GHz–300 GHz) , the last chapters of the book treat optical phenomenon, with EM waves in optical frequencies ($10^{14}-10^{15}$ Hz).

Chapter 1
Linear Systems and Electromagnetic Waves

Suppose a city gets a fresh water supply as a fountain or eruption from the ground. How will we use this water? We will build a tank to collect all this water, then use pipes to carry this water to houses, and then have taps at the end of pipes to get water. Now imagine instead of water, I have electromagnetic (EM) waves, say microwaves (electromagnetic waves with frequency 3–300 GHz). I have a supply of these microwaves, as radiation from electrons circulating a magnetic field. How do we collect this radiation? How do we channel this radiation? How do we then use this radiation in, say, a radio detection and ranging (RADAR) that detects flying objects, using microwaves or a microwave oven that cooks food using microwaves? As we see subsequently, we really collect this radiation in a tank called cavity and then use pipes called waveguides to take it to an antenna (the analog of tap), which directs these waves toward a flying object. How do we fill a cavity with microwaves? How do we model such electromagnetic waves? What equations govern their evolution in space and time. We study all this in this book. We model generation, storage, and transmission of electromagnetic waves as a linear system. We show that such a perspective of linear system lends great mathematical insight into the analysis of equations that govern the generation, storage, and transmission of electromagnetic waves. So what do we mean by a linear system? Let us have a look.

1.1 Linear Systems

Consider a mass M, tied by spring with spring constant k to a wall, as in Fig. 1.1. The displacement x of the mass from its equilibrium is related to force F as

$$M\ddot{x} + kx = F. \tag{1.1}$$

Using $\omega_0^2 = \frac{k}{M}$ and $u = \frac{F}{M}$, we get for $x_1 = \omega_0 x$ and $x_2 = \dot{x}$,

N. Khaneja, *Electromagnetic Waves and Optics*,
https://doi.org/10.1007/978-3-031-76144-7_1

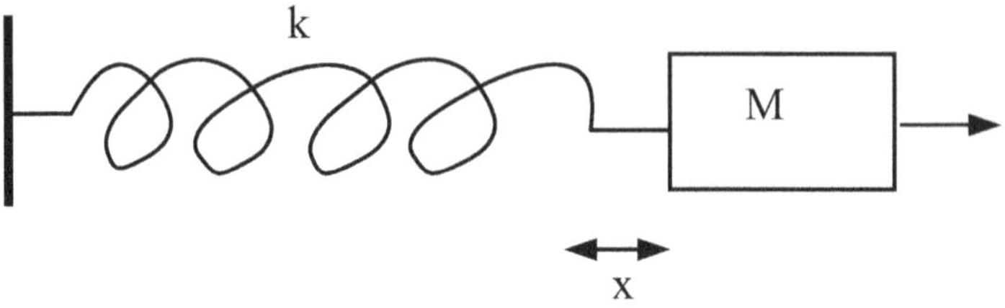

Fig. 1.1 A mass M tied to wall, by a spring with spring constant k, being pulled with force F. x measures displacement from equilibrium

$$\frac{d}{dt}\begin{bmatrix} x_1 \\ x_2 \end{bmatrix} = \underbrace{\begin{bmatrix} 0 & \omega_0 \\ -\omega_0 & 0 \end{bmatrix}}_{A}\begin{bmatrix} x_1 \\ x_2 \end{bmatrix} + \underbrace{\begin{bmatrix} 0 \\ 1 \end{bmatrix}}_{b} u. \tag{1.2}$$

If we denote $X = \begin{bmatrix} x_1 \\ x_2 \end{bmatrix}$, we can write the above system as

$$\frac{dX}{dt} = AX + bu. \tag{1.3}$$

In general, A is a $n \times n$ matrix, X, b are $n \times 1$ vector, and u is a time-varying input. We call Eq. (1.3) a *linear system.*

Without an input, the system takes the form

$$\dot{X} = AX. \tag{1.4}$$

Using $X(t + \Delta t) = (I + A\Delta t)X(t)$, we get $X(n\Delta t) = (I + A\Delta t)^n X(0)$; with $t = n\Delta t$, we get $X(t) = (I + \frac{At}{n})^n X(0)$. With $\Delta t \to 0$ and $n \to \infty$, we have $X(t) = \exp(At)X(0)$, where

$$\exp(At) = I + At + \frac{(At)^2}{2!} + \cdots + \frac{(At)^n}{n!} + \ldots. \tag{1.5}$$

Suppose u Eq. (1.3) is nonzero over an interval τ and $\tau + \Delta$. Then $X(\tau + \Delta t) = (I + A\Delta t)X(\tau) + bu(\tau)\Delta t$, and continuing, we get

$$\begin{aligned} X(t) &= \exp(A(t-\tau))X(\tau) + \exp(A(t-\tau))bu(\tau)\Delta t \\ &= \exp(At)X(0) + \exp(A(t-\tau))bu(\tau)\Delta t. \end{aligned}$$

and then in general for nonzero u, we have

$$X(t) = \exp(At)X(0) + \int_0^t \exp(A(t-\tau))bu(\tau)d\tau. \tag{1.6}$$

This is called the *variation of constant formula*. Although Eq. (1.6) solves Eq. (1.3), we show how we can further simplify.

To solve for evolution of the system Eq. (1.4), we observe that if the initial state $X(0)$ of the system is e, an eigenvector of A with eigenvalue λ, then the solution is simply $X(t) = \exp(\lambda t)e$. Simply substitute into Eq. (1.4). In general, we can decompose $X(0) = \sum_i \alpha_i(0)e_i$, where e_i are eigenvectors of A with eigenvalue λ_i. Then observe $X(t) = \sum_i \alpha_i(0)\exp(\lambda_i t)e_i$ is the solution as can be seen by substitution.

The solution is understood as simply writing the differential equation in terms of basis e_i with $X(t) = \sum_i \alpha_i(t)e_i$; then the vector equation gets expressed as n scalar equations

$$\dot{\alpha_i} = \lambda_i \alpha_i, \tag{1.7}$$

with solution $\alpha_i(t) = \exp(\lambda_i t)\alpha_i(0)$. In the presence of input as in Eq. (1.3), we can express $b = \sum_i b_i e_i$, and equations take the form

$$\dot{\alpha_i} = \lambda_i \alpha_i + b_i u(t), \tag{1.8}$$

with solution

$$\alpha_i(t) = \exp(\lambda_i t)\alpha_i(0) + b_i \int_0^t \exp(\lambda_i(t-\tau))u(\tau)d\tau. \tag{1.9}$$

With $\alpha_i(0) = 0$, we simply have $\alpha_i(t) = b_i \int_0^t \exp(\lambda_i(t-\tau))u(\tau)d\tau$. Of particular interest is the case b_i corresponding to imaginary eigenvalues $j\omega_i$. Then $\alpha_i(t) = b_i \int_0^t \exp(j\omega_i(t-\tau))u(\tau)d\tau$. If the input $u(t) = \exp(j\omega t)$, and we choose $\omega = \omega_k$, then all modes average only the mode k survives giving $\alpha_k(t) = b_k \exp(j\omega_k t)t$. We say we excite the system on resonance with mode k, which builds as shown. This is the main idea of this text. We identify modes and their frequencies and then drive the system at the frequency of a mode to build that mode. We say we pump energy into a mode.

For example, in Eq. (1.2), we have eigenvalues as $j\omega_0$ and $-j\omega_0$ with eigenvectors $e_1 = \begin{bmatrix} 1 \\ j \end{bmatrix}$ and $e_2 \begin{bmatrix} 1 \\ -j \end{bmatrix}$. We write input $b = \begin{bmatrix} 0 \\ 1 \end{bmatrix} = -\frac{i}{2}(e_1 - e_2)$. If input $u(t) = \exp(j\omega_0 t)$, the mode e_1 builds up. If input $u(t) = \cos(\omega_0 t) = \frac{\exp(j\omega_0 t)+\exp(-j\omega_0 t)}{2}$, both the modes e_1 and e_2 build up.

1.1.1 Steady-State Solution

Given the linear system

$$\frac{dX}{dt} = AX + bu, \tag{1.10}$$

where A is dissipative, i.e., eigenvalues of A have negative real part. Let $u = \exp(j\omega t)$. Then we can simply observe that $x = X_0 \exp(j\omega t)$ is a solution to Eq. (1.10), where X_0 is the solution to algebraic equation

$$j\omega X_0 = AX_0 + b. \tag{1.11}$$

or

$$\underbrace{(j\omega\mathbf{1} - A)}_{\mathcal{A}} X_0 = b. \tag{1.12}$$

Since A is dissipative, we have $\mathcal{A}$ invertible and we can solve for X_0. We call it steady-state solution because if y is the true solution, then $z = y - x$ satisfies

$$\frac{dz}{dt} = Az \tag{1.13}$$

with $z(T) = \exp(AT)z(0)$, since A is dissipative, $z(T) \to 0$ as $T \to \infty$ and $y(T) \to x(T)$. In physical problems, one encounters A that has purely imaginary eigenvalues, as in spring mass problem in (1.2). However, this is an ideal situation; in practice, there is a small amount of friction present, which gives dissipative A. Therefore, even in such ideal situations, it makes sense to talk about steady-state solution keeping in mind negligibly small dissipation.

Consider a scalar linear system

$$\dot{x} = -\alpha x + u. \tag{1.14}$$

For $s = j\omega$ and input $u = u(s)exp(st)$, the steady state is $x(s)\exp(st)$, where $x(s)$ satisfies the algebraic equation

$$(s + \alpha)x(s) = u(s), \tag{1.15}$$

with

$$G(s) = \frac{x(s)}{u(s)} = \frac{1}{s + \alpha}. \tag{1.16}$$

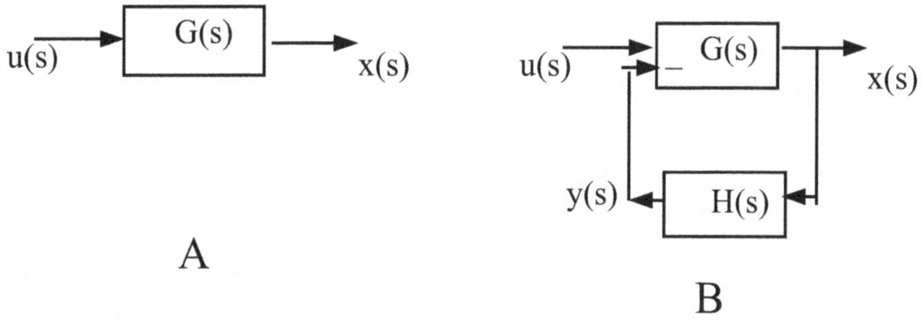

Fig. 1.2 Linear systems represented by transfer functions

$G(s)$ is called the input-output transfer function, which is the frequency response of the system as shown in Fig. 1.2a.

Now let us take $x(t)$ and feed it as input to another system

$$\dot{y} = -\beta y + x(t), \tag{1.17}$$

then

$$H(s) = \frac{y(s)}{x(s)} = \frac{1}{s + \beta}. \tag{1.18}$$

Now if we take y and feed it back to x with a negative sign

$$\dot{x} = -\alpha x + u - y, \tag{1.19}$$

as shown in Fig. 1.2b, then

$$x(s) = G(s)(u(s) - y(s)) = G(u(s) - Hx(s)), \tag{1.20}$$

or

$$C(s) = \frac{x(s)}{u(s)} = \frac{G(s)}{1 + G(s)H(s)}. \tag{1.21}$$

$C(s)$ is called the feedback or closed-loop transfer function.

1.2 Maxwell Equations

Why do we care about linear systems? Because we want to study electromagnetic waves [3, 4, 10, 12, 15]. We know electromagnetic waves everyday from radio and television (TV) broadcast, and light and color around us. These waves propagate as oscillating electric and magnetic fields. The equation governing their evolution are the Maxwell equations. Electromagnetic waves are space- and time-varying electric and magnetic fields related by Maxwell equations as in

$$\mu \frac{\partial H}{\partial t} = -\nabla \times E \tag{1.22}$$

$$\epsilon \frac{\partial E}{\partial t} = \nabla \times H - J(r, t), \tag{1.23}$$

where each spatial location r indices a vector entry in linear system as in Eq. (2.16). The vector entries are E (electric) and H (magnetic) fields. The goal is to calculate $E(r, t)$ and $H(r, t)$ resulting from an current input $J(r, t)$. That is, solve a linear system for an input $u(t)$. It means we have to find modes of the system and the corresponding eigenvalues. In other words, we have to solve for $Ax = \lambda x$ in Eq. (1.4), i.e., solve for

$$-\frac{1}{\mu} \nabla \times E = \lambda H, \tag{1.24}$$

$$\frac{1}{\epsilon} \nabla \times H = \lambda E. \tag{1.25}$$

Of particular interest is when $\lambda = j\omega$, imaginary eigenvalues in which case, the eigenvectors are as follows: For three vectors $k = |k|\hat{k}$, with $\hat{x}$, $\hat{y}$, $\hat{k}$ forming a right-handed system. For $\frac{E_0}{H_0} = \eta = \sqrt{\frac{\mu_0}{\sigma_0}}$, we have

$$E = E_0 \exp(-jk \cdot r)\hat{x} = E_0 \exp(-jk_x x + k_y y + k_z z)\hat{x}, \tag{1.26}$$

$$H = H_0 \exp(-jk \cdot r)\hat{y} = H_0 \exp(-jk_x x + k_y y + k_z z)\hat{y}. \tag{1.27}$$

is an eigenvector with $c|k| = \omega$. Thus, eigenvectors are the Fourier modes.

Now we want to drive Maxwell equations with current sources $J(r, t)$. This current source is an antenna. They can in fact be simplified to $J(r)u(t)$. Think of $J(r)$ as b in linear system. These sources are localized and can be expanded into a Fourier basis, but these Fourier bases are precisely the eigenfunctions of Maxwell equation with imaginary eigenvalues. Then if we choose $u(t) = \exp(j\omega t)$, then as described above, we will excite the modes with frequency ω. The antenna puts energy into the Fourier modes whose frequency is same as the frequency of excitation of the antenna. These modes get energy, while other modes do not, and we can detect these modes very far from the antenna because they are delocalized, and hence we can communicate to someone very far because we pump energy into a mode that is shared simultaneously between the transmitter and receiver. In summary, we care about eigenfunctions corresponding to imaginary eigenvalues as their natural evolution constitutes traveling waves, and we communicate by pumping energy in these modes.

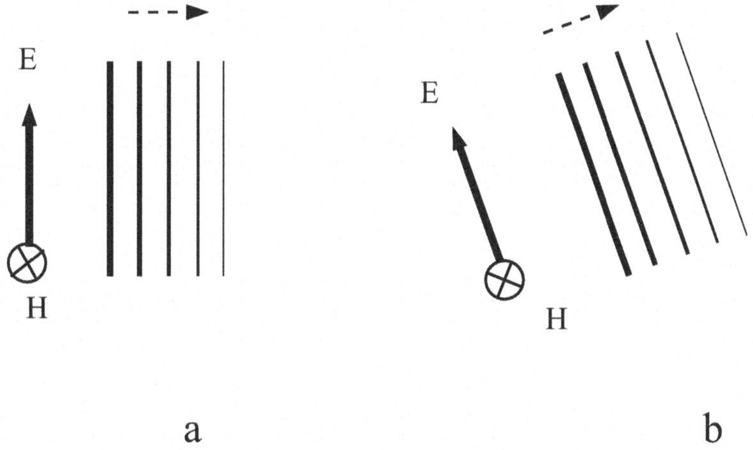

Fig. 1.3 (**a**) and (**b**) EM waves propagating along x and $\hat{k}$ directions

Till now, we discussed Maxwell equations in free space. The eigenmodes with imaginary eigenvalues are Fourier modes. The evolution of these modes is

$$E = E_0 \exp(j(\omega t - k \cdot r))\hat{x}, \tag{1.28}$$

$$H = H_0 \exp(j(\omega t - k \cdot r))\hat{y}. \tag{1.29}$$

This is an EM wave propagating in the k direction (Fig. 1.3).

Another very interesting situation is when Maxwell equations are confined in a cavity. A cavity is a free space, enclosed with conducting walls. EM waves do not propagate inside the conductor, leading to zero tangential electric fields and zero normal magnetic fields at boundaries. We can solve the eigenvectors of Maxwell equations with these boundary conditions to get what are called modes of the cavities. These modes, say

$$\phi_1, \phi_2, \ldots, \phi_k, \ldots,$$

have eigenvalues

$$j\omega_1, j\omega_2, \ldots, j\omega_k, \ldots.$$

When we excite the cavity, with a current source, again we can expand the localized source into cavity eigenfunctions. Figure 1.4 shows a current source as wire inserted in the cavity. If the source has current $\exp(j\omega_k t)$, we pump energy into mode ϕ_k, and the mode ϕ_k will build up and all cavity will get filled with mode ϕ_k. We show how to fill cavities with microwaves and channel them through pipes (waveguides) for many applications. Electromagnetic waves in 30 kHz to 3 GHz are called radio waves. 3–300 GHz are called microwaves. 10^{12}–10^{14} Hz EM waves are called

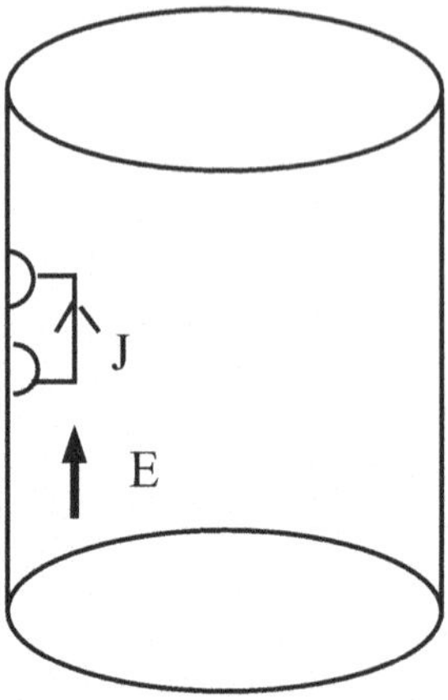

Fig. 1.4 (**a**) Exciting modes of cavity by putting in a wire carrying oscillating current

infrared light and $10^{14}-10^{15}$ Hz visible light, and so on. Each frequency type has its own applications. Let us look at some of these applications before we jump into the main content of this book.

1.3 Applications of Electromagnetic Waves

1.3.1 Optical Fibers

These days, long-range communication between cities and countries is done using fiber optic communication. The light used is infrared frequency 180−330 THz. This light is carried by an optical fiber [27] which is a cylindrical dielectric rod made of silica, surrounded by a dielectric shell called cladding as in Fig. 1.5a. Light moves in the core. The refractive index n_1 of core is greater than the refractive index n_2 of surrounding air. When light tries to leave the core as in Fig. 1.5b, it gets totally internally reflected for small θ. This way, fiber confines light and light travels far.

1.3.2 Radio and Television Broadcast

The most traditional application of electromagnetic waves is radio and television broadcast. Radio transmission is in low radio frequency 1−2 MHz, amplitude-modulated (AM) radio, or very high frequency (VHF) (30−300 MHz) for frequency-modulated (FM) radio. AM broadcast over long distance is done over short wave (high frequency (HF) (3−30 MHz). Radio waves in the shortwave band can be reflected or refracted from a layer of electrically charged atoms in the

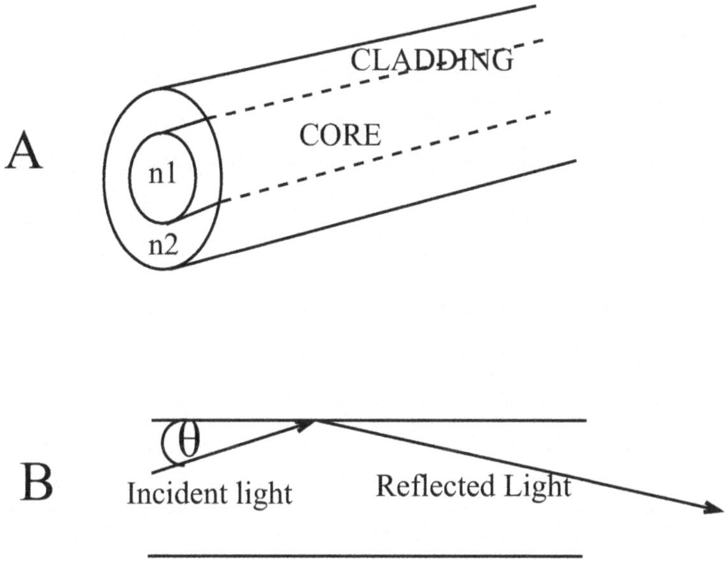

Fig. 1.5 (**a**) Optical fiber, with core and cladding. (**b**) Light getting totally internally reflected when it tries to leave the fiber

atmosphere called the ionosphere. Therefore, short waves directed at an angle into the sky can be reflected back to Earth at great distances, beyond the horizon. This is called skywave or "skip" propagation [12, 25]. Thus, shortwave radio can be used for very long-distance communication, in contrast to radio waves of higher frequency which travel in straight lines (line-of-sight propagation) and are limited by the visual horizon, about 64 km (40 miles). Television broadcast is done in VHF. Cell phones we use every day work in ultrahigh frequency (UHF; 300 MHz to 3 GHz). This band is also used for television broadcasting, satellite communication including global positioning system (GPS), personal radio services including Wi-Fi and Bluetooth, walkie-talkies, cordless phones, and numerous other applications [18, 23].

1.3.3 Satellite Communications

We all have dish TV at home, where signal is received from a satellite [20]. The signal is first transmitted to satellite using an uplink, which operates in Ku microwave band (12−18 GHz) frequency. The is downconverted and broadcast in C-Band (4−8 GHz). The signals are received via an outdoor parabolic antenna, commonly referred to as a satellite dish.

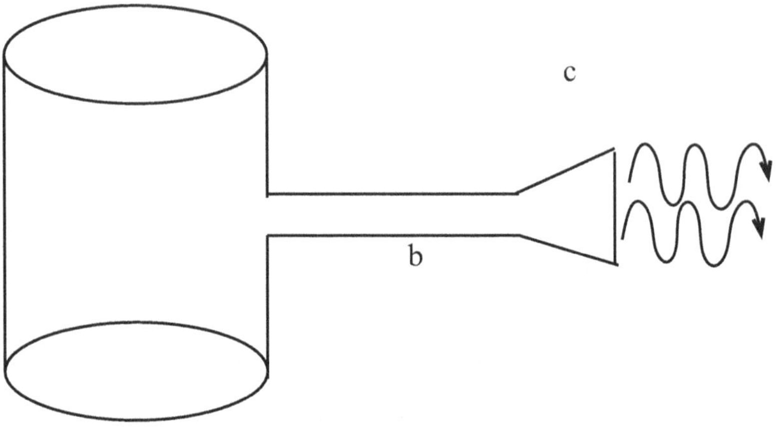

Fig. 1.6 Schematic of a RADAR, including oscillator cavity, waveguide, and Horn antenna

1.3.4 Microwave Links

Pre-fiber optic days, line of sight microwave links provided telephone connectivity between cities. Typical frequencies used were less than 10 GHz. These days, terrestrial microwave relay links are used in telecommunication networks, including cellular networks linking base transceiver station (BTS), base station controller (BSC), and mobile switching center (MSC). Frequencies from $10 to 80$ GHz are used.

1.3.5 Radar

One of the most prominent applications of microwaves is the microwave radar [26, 31]. Radar is a detection system that uses microwaves to determine the range, angle, or velocity of objects. It can be used to detect aircraft, ships, spacecraft, guided missiles, motor vehicles, weather formations, and terrain. A radar system consists of a transmitter, producing electromagnetic waves in the microwave domain, a transmitting antenna, a receiving antenna (often the same antenna is used for transmitting and receiving), and a receiver and processor to determine the properties of the object(s). Microwaves (pulsed or continuous) from the transmitter reflect off the object and return to the receiver, giving information about the object's location and speed. Figure 1.6 shows a schematic of radar where an oscillator cavity (magnetron and klystron) labeled a produces microwaves which are channeled by waveguides, labeled b to Horn antenna, labeled c.

The modern uses of radar are highly diverse, including air and terrestrial traffic control, air-defense systems, antimissile systems, marine radars to locate landmarks and other ships, aircraft anti-collision systems, ocean surveillance systems, outer space surveillance and rendezvous systems, meteorological precipitation monitoring, altimetry and flight control systems, guided missile target locating systems, and ground-penetrating radar for geological observations. Radar frequency ranges from 3 to 30 GHz.

1.3.6 Remote Sensing

Electromagnetic radiation in the microwave wavelength region is used in remote sensing to provide useful information about the Earth's atmosphere, land, and ocean. A radar, mounted on a satellite, transmits microwave pulses toward the earth surface. The microwave energy scattered back to the spacecraft is measured. Remote sensing makes use of the radar principle to form an image by utilizing the time delay of the backscattered signals. In real aperture radar imaging, the ground resolution is limited by the size of the microwave beam sent out from the antenna. Finer details on the ground can be resolved by using a narrower beam. The beamwidth is inversely proportional to the size of the antenna, i.e., the longer the antenna, the narrower the beam. The microwave beam sent out by the antenna illuminates an area on the ground (known as the antenna's "footprint"). In radar imaging, the recorded signal strength depends on the microwave energy, backscattered from the ground targets, inside this footprint. Increasing the length of the antenna will decrease the width of the footprint. It is not feasible for a spacecraft to carry a very long antenna, which is required for high-resolution imaging of the earth surface. To overcome this limitation, the principle of synthetic aperture radar (SAR) is used. SAR capitalizes on the motion of the space craft to emulate a large antenna from the small antenna it actually carries onboard. When microwaves strike a surface, the proportion of energy scattered back to the sensor depends on many factors: Physical factors such as the dielectric constant of the surface materials which also depends strongly on the moisture content; geometric factors such as surface roughness, slopes, and orientation of the objects relative to the radar beam direction; the types of land cover (soil, vegetation, or man-made objects). Microwave frequency, polarization, and incident angle. The ability of microwave to penetrate clouds, precipitation, or land surface cover depends on its frequency. Generally, the penetration power increases for longer wavelength (lower frequency). The SAR backscattered intensity generally increases with surface roughness. However, "roughness" is a relative quantity. Whether a surface is considered rough or not depends on the length scale of the measuring instrument. If a meter rule is used to measure surface roughness, then any surface fluctuation of the order of 1 cm or less will be considered smooth. On the other hand, if a surface is examined under a microscope, then a fluctuation of the order of a fraction of a millimeter is considered very rough. In SAR imaging, the reference length scale for surface roughness is the wavelength of the microwave. If the surface fluctuation is less than the microwave wavelength, then the surface is considered smooth. For example, little radiation is backscattered from a surface with a fluctuation of the order of 5 cm if an L-band (15–30 cm wavelength, 1 GHz frequency) SAR is used and the surface will appear dark. However, the same surface will appear bright due to increased backscattering in an X-band (2.4–3.8 cm wavelength, 10 GHz frequency) SAR image. The C-band is useful for imaging ocean and ice features. However, it also finds numerous land applications. The L-band has a longer wavelength and is more penetrating than the C-band. Hence, it is more useful in forest and vegetation study as it is able to penetrate deeper into the vegetation canopy.

1.3.7 Radio Astronomy

Radio astronomy is a subfield of astronomy that studies celestial objects at radio frequencies [5, 16]. For example, radiation at 21 cm, 1428 MHz frequency is emitted from galactic hydrogen due to hyperfine splitting of its orbital. We can detect this radiation with high angular precision, which gives the angular location of the distant galaxy. The Doppler shift in the frequency gives an estimate of how fast the galaxy is receding from us. The angular precision is measured by the ratio $\frac{\lambda}{D}$, where D is the diameter of the antenna. With a large wavelength, we need large diameter D. This is made feasible by having an array of such antennas a technique called synthetic aperture radar similar to remote sensing.

1.3.8 Microwave Heating

Water in the food absorbs microwaves around 2.450 GHz. At this frequency, transitions between rotational states are induced in the water molecule. The molecule in excited rotational state dissipates energy into heat. This forms the basis of what is called a microwave heating in microwave oven. Like radar in a microwave oven, we have a cavity to produce microwaves, which are carried by a short waveguide to the main cavity which is filled with microwaves to heat the food.

1.3.9 Spectroscopy

One of the big applications of electromagnetic waves is spectroscopy. Using electromagnetic waves, we can probe the structure of atoms and molecules. Different types of spectroscopies correspond to different frequency electromagnetic waves. Radio frequency electromagnetic waves can induce transition between the spin states of atomic nucleus placed in a magnetic field B_0. The nuclear spin is like a charged spinning top which carries a magnetic moment μ. The difference in the energy of spin (magnetic moment) oriented along and opposite to the magnetic field is $E = 2\mu B_0$. Writing this energy in frequency $E = \hbar\omega_0$, we find that ω_0 for hydrogen at a field of 10 T corresponds to around 400 MHz. Thus, radio frequency EM can probe nuclear spins, and this field of spectroscopy is called nuclear magnetic resonance (NMR) [1] spectroscopy. Electron spin has μ around 10^3 times higher than that of hydrogen nucleus, and we can probe electron spin at around the same B_0 field strength by GHz radiation. This field is called electron paramagnetic resonance (EPR). The energy difference in rotational states of molecules corresponds to GHz frequency, and the resulting spectroscopy is called microwave spectroscopy. The energy difference in the vibrational states of molecules correspond to THz frequency, and the resulting spectroscopy is called vibrational spectroscopy. The transition of electrons between different orbitals corresponds to optical 10^{15} Hz frequency, and the resulting spectroscopy is called optical spectroscopy.

1.4 Problems

1. The relative permittivity of a medium is 1.4. Find the velocity of light in the medium.
2. The frequency of EM wave is 1 GHz. Find its wavelength.
3. What is the frequency of 1 mm microwave?
4. The relative permittivity of a medium is 1.5. Find the wavelength of 100 GHz microwave.
5. How long is a 1 GHz broad microwave packet in time and space?

Chapter 2
Electromagnetic Waves and Propagation

2.1 Maxwell Equations

Our starting point is the electric field due to a point charge q_0 which is

$$E = \frac{1}{4\pi\epsilon_0}\frac{q_0}{r^2}\hat{r}. \tag{2.1}$$

where ϵ_0 is the permittivity of free space with $\frac{1}{4\pi\epsilon_0} = 9 \times 10^9$ SI units. By use of the Stokes' theorem, the above equation can be written as

$$\nabla E = \frac{\rho}{\epsilon_0}. \tag{2.2}$$

where ∇E is the divergence of E and ρ charge density. Equation (2.2) constitutes our first Maxwell equation.

If we bring a test charge q, in the presence of electric field E, it feels a force $F = qE$. In addition to electric fields, we also have magnetic fields. They arise due to moving charges. When charge q_0 moves, it produces a magnetic field. Consider a wire carrying current i_0, then the magnetic field at location O, due to infinitesimal current element, is given by Biot-Savart law [9, 21], as shown in Fig. 2.1a.

$$B = \frac{\mu_0}{4\pi}\frac{i_0 dl \times r}{r^3} = \frac{\mu_0}{4\pi}\frac{i_0 dl \sin\theta}{r^2}\hat{\phi}. \tag{2.3}$$

where ϕ is the unit vector at O going inside the page.

As can be seen

$$\nabla \cdot B = 0. \tag{2.4}$$

N. Khaneja, *Electromagnetic Waves and Optics*,
https://doi.org/10.1007/978-3-031-76144-7_2

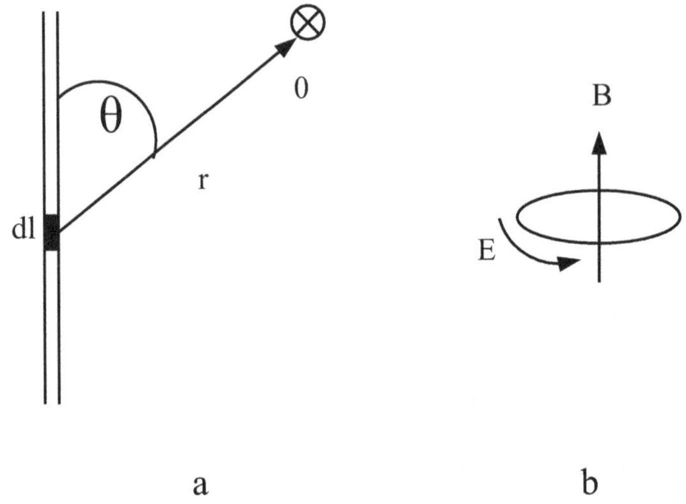

Fig. 2.1 (**a**) Magnetic field at point O going inside the page caused due to infinitesimal current element. (**b**) Change of magnetic flux through a loop inducing tangential electric field

If we integrate over the whole wire, we get the net field at location O as

$$B = \frac{\mu_0}{2\pi}\frac{i_0}{r}\hat{\phi}. \tag{2.5}$$

If we have current density J at location x and we come infinitely close to this, then the density looks infinitely long and we can write B as in Eq. (2.5). If A is the area of the current density, we can write $B(2\pi r) = (\nabla \times B)A$ and $i_0 = JA$, which from Eq. (2.5) gives

$$\nabla \times B = \mu_0 J, \tag{2.6}$$

or

$$\nabla \times H = J. \tag{2.7}$$

Electric field E induces a dipole per unit volume given by the displacement $D = \epsilon E$. Then $\frac{\partial D}{\partial t}$ tantamounts to a current density and Eq. (2.7) generalizes to

$$\nabla \times H = J + \frac{\partial D}{\partial t}. \tag{2.8}$$

Finally, consider a loop of wire (of radius r and area A) with magnetic field B passing through it as shown in Fig. 2.1b. This results in flux $\Phi = BA$ through the loop. Change of flux induces a tangential electric field in the loop whose line integral is given by

$$E(2\pi r) = -\frac{\partial \phi}{\partial t}. \tag{2.9}$$

As we shrink the loop, we get for infinitesimal loop $E(2\pi r) = (\nabla \times E)A$, which gives the famous Faraday's law

$$\nabla \times E = -\frac{\partial B}{\partial t}. \tag{2.10}$$

Putting equations (2.2), (2.4), (2.8), and (2.10) together, we get the famous Maxwell equations [19].

2.2 Wave Equation and Linear Systems

The Maxwell equations are [13]

$$\nabla \cdot E = \frac{\rho(x,t)}{\epsilon_0}, \tag{2.11}$$

$$\nabla \cdot B = 0, \tag{2.12}$$

$$\frac{\partial B}{\partial t} = -\nabla \times E, \tag{2.13}$$

$$\frac{\partial D}{\partial t} = \nabla \times H - J(x,t). \tag{2.14}$$

Suppose we start with E and B satisfying Eqs. (2.11 and 2.12). Then evolution under Eqs. (2.13 and 2.14) ensures they are satisfied for all t. All we have to take notice is the fact that the divergence of curl is zero and the equation of continuity

$$\frac{\partial \rho(x,t)}{\partial t} + \nabla \cdot J = 0. \tag{2.15}$$

So we focus on Eqs. (2.13 and 2.14). These equations constitute, what we call a linear system. In a linear system, we call the state of the system X, which evolves according to the equation

$$\dot{X} = AX + bu(t), \tag{2.16}$$

where X is say an $n \times 1$ vector, A an $n \times n$ matrix, b an $n \times 1$ input vector, and $u(t)$ input. Without an input, the system takes the form

$$\dot{X} = AX. \tag{2.17}$$

To solve for evolution of this system, we observe if the initial state $X(0)$ of the system is e, an eigenvector of A, with eigenvalue λ, then the solution is simply $X(t) = \exp(\lambda t)e$. Simply substitute it in Eq. (2.17). In general, we can decompose $X(0) = \sum_i \alpha_i(0)e_i$, where e_i are eigenvectors of A with eigenvalue λ_i. Then observe that $X(t) = \sum_i \alpha_i(0)\exp(\lambda_i t)e_i$ is the solution as can be seen by substitution.

The solution is understood as simply writing the differential equation in terms of basis e_i with $X(t) = \sum_i \alpha_i(t)e_i$, then the vector equation gets expressed as n scalar equations

$$\dot{\alpha_i} = \lambda_i \alpha_i, \tag{2.18}$$

with solution $\alpha_i(t) = \exp(\lambda_i t)\alpha_i(0)$. In the presence of input as in Eq. (2.16), we can express $b = \sum_i b_i e_i$, and equations take the form

$$\dot{\alpha_i} = \lambda_i \alpha_i + b_i u(t), \tag{2.19}$$

with solution

$$\alpha_i(t) = \exp(\lambda_i t)\alpha_i(0) + b_i \int_0^t \exp(\lambda_i(t-\tau))u(\tau)d\tau. \tag{2.20}$$

With $\alpha_i(0) = 0$, we simply have $\alpha_i(t) = b_i \int_0^t \exp(\lambda_i(t-\tau))u(\tau)d\tau$. Of particular interest is the case when λ_i corresponds to imaginary eigenvalues $j\omega_i$. Then $\alpha_i(t) = b_i \int_0^t \exp(j\omega_i(t-\tau))u(\tau)d\tau$. If the input $u(t) = \exp(j\omega t)$ and we choose $\omega = \omega_k$, then all modes average only the mode k survives giving $\alpha_k(t) = b_k \exp(j\omega_k t)t$. We say we excite the system on resonance, with mode k, which builds as shown. This is the main idea of this text. We identify modes and their frequencies and then drive the system at the frequency of a mode to build that mode. We say we pump energy into a mode. Let us now go to Maxwell equations in

$$\mu\frac{\partial H}{\partial t} = -\nabla \times E, \tag{2.21}$$

$$\epsilon\frac{\partial E}{\partial t} = \nabla \times H - J(r,t), \tag{2.22}$$

where each spatial location r indices a vector entry in linear system as in Eq. (2.16). The vector entries are E and B fields. The goal is to calculate $E(r,t)$ and $B(r,t)$ resulting from an current input $J(r,t)$. That is, solve a linear system for an input $u(t)$. It means we have to find modes of the system and the corresponding eigenvalues. In other words, we have to solve for $Ax = \lambda x$ in Eq. (2.17), i.e., solve for

$$-\frac{1}{\mu}\nabla \times E = \lambda H, \tag{2.23}$$

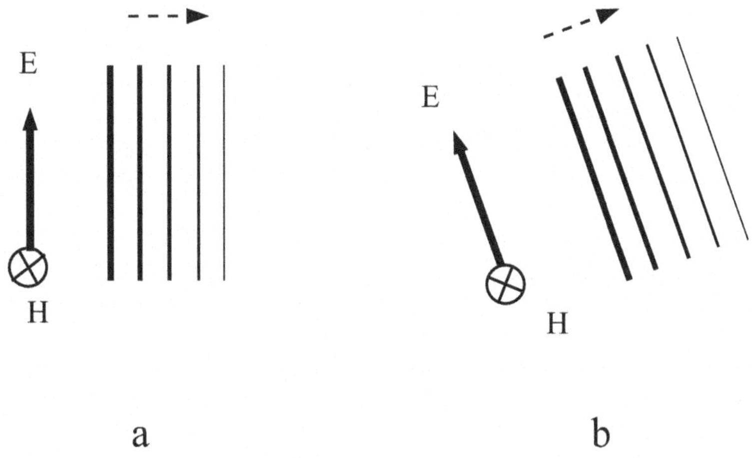

Fig. 2.2 (**a**) and (**b**) EM waves propagating along x and $\hat{k}$ directions

$$\frac{1}{\epsilon}\nabla \times H = \lambda E, \tag{2.24}$$

with $E = E_o \exp(-kx)\ \hat{y}$ and $H = H_o \exp(-kx)\ \hat{z}$, where $\hat{x}$, $\hat{y}$, $\hat{z}$ are standard unit vectors, we get

$$\frac{k}{\mu} E_o = \lambda H_o, \tag{2.25}$$

$$\frac{k}{\epsilon} H_o = \lambda E_o, \tag{2.26}$$

which gives $ck = \pm\lambda$, where $c^2 = \frac{1}{\mu\epsilon}$ and $\frac{E_o}{H_o} = \pm\eta$, where $\eta = \sqrt{\frac{\mu}{\epsilon}}$. Of particular interest is when $\lambda = j\omega$ is imaginary, then $E = E_o \exp(-jkz)\ \hat{x}$ and $H = H_o \exp(-jkz)\ \hat{y}$ with $ck = \omega$. The modes evolve as $E = E_o \exp(-j(kx - \omega t))\ \hat{y}$ and $H = H_o \exp(-j(kx - \omega t))\ \hat{z}$. This constitutes a traveling wave along x direction, as shown in Fig. 2.2a. There is another eigenvector corresponding to $ck = -\omega$, which evolves as $E = E_o \exp(j(kx + \omega t))\ \hat{y}$ and $H = -H_o \exp(j(kx + \omega t))\ \hat{z}$ and constitutes a traveling wave along $-x$ direction.

Let $\hat{\mathbf{x}}, \hat{\mathbf{y}}, \hat{\mathbf{z}}$ be some unit vectors forming a right-handed coordinate system such that $\hat{\mathbf{x}} = \frac{k}{|k|}$ with $k = (k_x, k_y, k_z)$ and $E = E_o \exp(-j\ k_x x + k_y y + k_z z)\ \hat{\mathbf{y}}$ also written as $E = E_o \exp(-jk \cdot r)\ \hat{\mathbf{y}}$ and $H = H_o \exp(-jk \cdot r)\ \hat{\mathbf{z}}$ are eigenvectors corresponding to eigenvalue $j\omega$ and $E = E_o \exp(-j(k \cdot r - \omega t))\ \hat{\mathbf{y}}$ and $H = H_o \exp(-j(k \cdot r - \omega t))\ \hat{\mathbf{z}}$ is a traveling wave along k direction as shown in Fig. 2.2b, and there is a corresponding wave $E = E_o \exp(j(k \cdot r + \omega t))\ \hat{\mathbf{x}}$ and $H = H_o \exp(j(k \cdot r + \omega t))\ \hat{\mathbf{y}}$ traveling along $-k$ direction.

What have we learnt? The eigenfunctions corresponding to imaginary eigenvalues are oscillatory in time, and their evolution constitutes traveling waves. Eventually, we want to drive Maxwell equations with current sources $J(r, t)$. These are antenna's. They can in fact be simplified to $J(r)u(t)$. Think of $J(r)$ as b in linear system. These sources are localized and can be expanded into a Fourier basis, but these Fourier bases are precisely the eigenfunctions of Maxwell equation, with imaginary eigenvalues. Then, if we choose $u(t) = \exp(j\omega t)$ as described above, we will excite the modes with frequency ω. The antenna puts energy into traveling wave modes, whose frequency is the same as the frequency of excitation of the antenna. These modes get energy and become much bigger than other modes, and we can detect these modes very far from the antenna because they are delocalized, and hence we can communicate to someone very far because we pump energy into a mode that is shared simultaneously between the transmitter and receiver. In subsequent chapters, we present other elegant techniques for solving for electric and magnetic fields arising due to antenna excitation. In summary, we care about eigenfunctions corresponding to imaginary eigenvalues as their natural evolution constitutes traveling waves, and we communicate by pumping energy in these modes.

2.3 Wave Equation and Media Boundary Conditions

We have talked about traveling waves $E = E_0 \exp(-j(k \cdot r - \omega t)\hat{x}$ and $H = H_0 \exp(-j(k \cdot r - \omega t)\hat{y}$, with $\frac{E_0}{H_0} = \eta = \sqrt{\frac{\mu}{\epsilon}}$. This is transmission in free space. How about transmission across a media interface? Consider the scenario as depicted in Fig. 2.3. We have a transmitted wave in region I that arrives at the media interface, which is say the y−z plane. Region I is $x < 0$ and region 2 is $x > 0$. The permittivity of I is ϵ_1 and II is ϵ_2. The incident wave cannot continue unaltered into region II as η for region II is different. We will need to alter E_0 and H_0. We do this through a reflected wave and transmitted wave. We say there is a reflected wave that goes back in region I and transmitted wave that proceeds in region II. This is depicted in Fig. 2.3.

If we denote the spatial dependence of incident, reflected, and transmitted waves as $\exp(-jk_I \cdot r)$, $\exp(-jk_r \cdot r)$ and $\exp(-jk_t \cdot r)$, then our first claim is that k_I, k_r, k_t are in the same plain. Without loss of generality (W.L.O.G), assume k_I in the x, z plain. Then first note that tangential and normal components of E and H are the same at the interface. This follows from Fig. 2.4. In Fig. 2.4a,

$$E_2 - E_1 = (\nabla \times E)_\perp \Delta x, \tag{2.27}$$

where $\nabla \times E$ is the curl and $(\nabla \times E)_\perp$ its component perpendicular to the plane. The wave arrangement of incident, transmitted, and reflected waves is an eigenfunction of the Maxwell equation satisfying

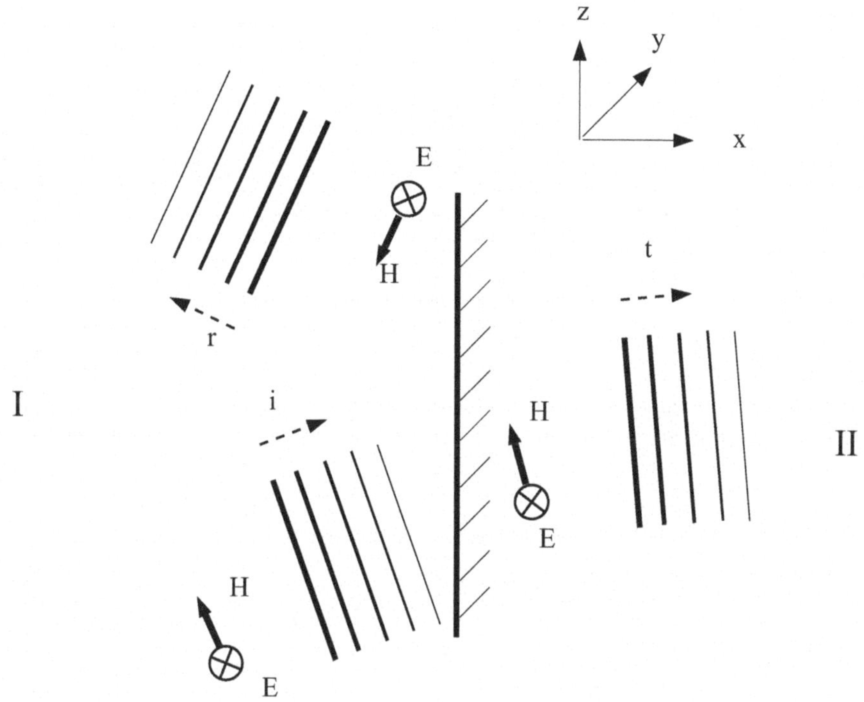

Fig. 2.3 Incident, transmitted, and reflected waves across a media interface between media *I* and *II*

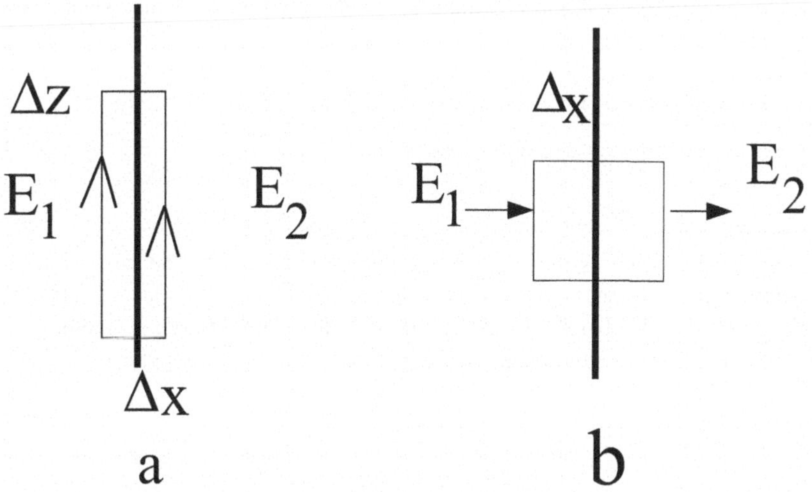

Fig. 2.4 (**a**) and (**b**) Tangential and normal components of the electric field across a media interface

$$-\frac{1}{\mu}\nabla \times E = j\omega H \tag{2.28}$$

$$\frac{1}{\epsilon}\nabla \times H = j\omega E \tag{2.29}$$

Hence, curl is finite and therefore in Eq. (2.27) as $\Delta x \rightarrow 0$ we get $E_2 = E_1$. Furthermore, from Eqs. (2.28 and 2.29), we find that $\nabla \cdot E = 0$ and $\nabla \cdot H = 0$. Therefore, from Fig. 2.4b, we get for $\Delta x \rightarrow$

$$E_2 - E_1 = (\nabla \cdot E)\Delta z, \tag{2.30}$$

and since $\nabla \cdot E$ is finite, we have $E_1 = E_2$. The same is true for H, and as we get across the boundary the tangential and normal components of E and H are the same. Let us compare y component of E; we get

$$\begin{aligned} &E_{io}\exp(-j(k_{ix}x + k_{iz}z)) + E_{ro}\exp(-j(k_{rx}x + k_{ry}y + k_{rz}z)) \\ &\quad = E_{to}\exp(-j(k_{tx}x + k_{ty}y + k_{tz}z)). \end{aligned} \tag{2.31}$$

Let us only vary y to get

$$E_{io} + E_{ro}\exp(-jk_{ry}y) = E_{to}\exp(-jk_{ty}y). \tag{2.32}$$

Differentiating, we get $k_{ty} = k_{ry}$, which further implies $k_{ty} = k_{ry} = 0$, which means transmitted and reflected waves are in the same plane. Then writing Eq. (2.31), we get from Fig. 2.5

$$\begin{aligned} &E_{io}\exp(-jk_i(\sin\theta_i z + \cos\theta_i x)) + E_{ro}\exp(-jk_i(\sin\theta_r z - \cos\theta_r x)) \\ &\quad = E_{to}\exp(-jk_t(\sin\theta_t z + \cos\theta_t x)). \end{aligned} \tag{2.33}$$

Again only vary z to conclude $\sin\theta_i = \sin\theta_r$ and $\frac{\sin\theta t}{\sin\theta i} = \frac{k_i}{k_t} = \sqrt{\frac{\epsilon_1}{\epsilon_2}}$. That is, the angle of incidence is the same as the angle of reflection. The ratio of sine of transmitted and incident angles obeys which is known as Snell's law of refraction. The transmitted wave bends inward as it moves into a dense (larger ϵ) media.

Now equating tangential components of E and H at the origin gives

$$E_{io} + E_{ro} = E_{to}, \tag{2.34}$$

$$(E_{io} - E_{ro})\frac{\cos\theta_i}{\eta_1} = E_{to}\frac{\cos\theta_t}{\eta_2}, \tag{2.35}$$

which gives

$$(E_{io} + E_{ro}) = E_{to}, \tag{2.36}$$

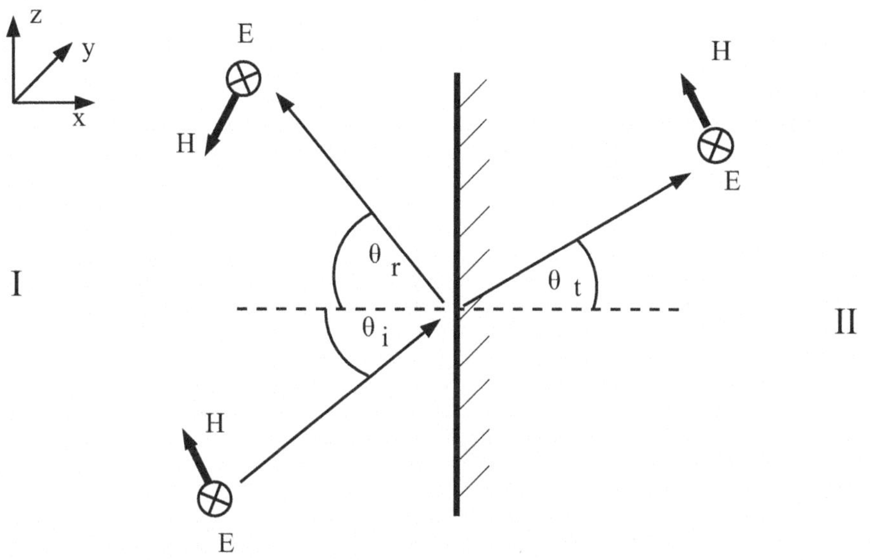

Fig. 2.5 Incident, transmitted, and reflected waves at the media interface of medium I and II. E field is parallel to the interface

$$(E_{io} - E_{ro}) = \frac{\eta_1 \cos\theta_t}{\eta_2 \cos\theta_i} E_{to}, \tag{2.37}$$

with $a = \frac{\eta_1}{\eta_2}$ and $b = \frac{\cos\theta_t}{\cos\theta_i}$; we have $E_{to} = \frac{2E_{io}}{1+ab}$ and $E_{ro} = \frac{1-ab}{1+ab} E_{io}$, where b is calculated from $\frac{\sin\theta_t}{\sin\theta_i} = a^{-1}$.

For normal incidence, with $\theta_i = \theta_t = 0$, we get $b = 1$ and $E_{ro} = \frac{1-a}{1+a} E_{io}$.

We discussed the case when incident wave has E tangential to the interface while H is incident at an angle. The second scenario is when incident wave has H tangential to the interface while E is incident at an angle. This is shown in Fig. 2.6.

For this case, we get

$$(E_{io} + E_{ro}) = \frac{\eta_1}{\eta_2} E_{to}, \tag{2.38}$$

$$(E_{io} - E_{ro}) \cos\theta_i = E_{to} \cos\theta_t, \tag{2.39}$$

which gives $E_{to} = \frac{2E_{io}}{a+b}$ and $E_{ro} = \frac{(a-b)E_{i0}}{a+b}$.

For normal incidence with $\theta_i = \theta_t = 0$, we get $b = 1$ and $E_{ro} = \frac{a-1}{a+1} E_{io}$.

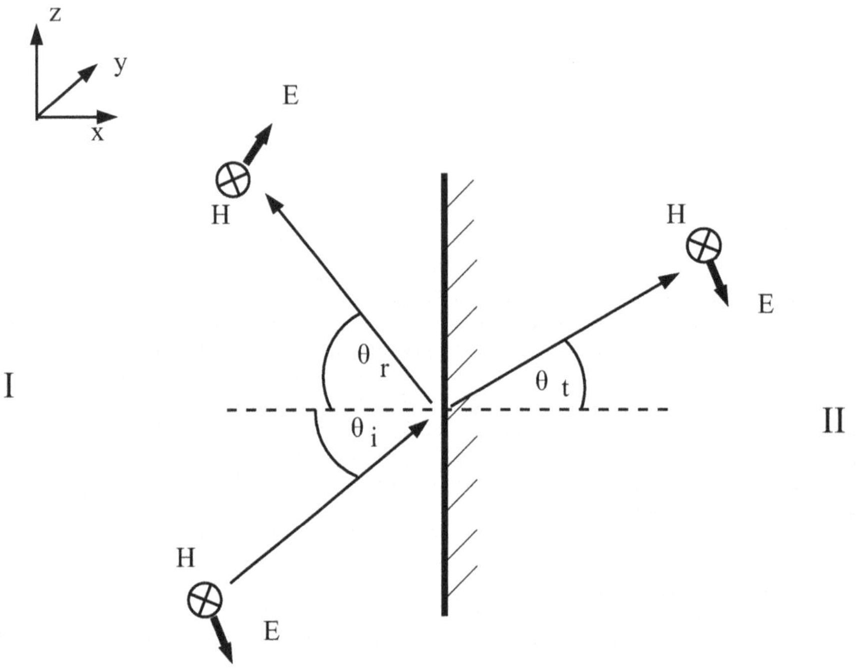

Fig. 2.6 Incident, transmitted, and reflected waves at the media interface of medium I and II. H field is parallel to the interface

2.4 Conductor Interfaces

$$-\nabla \times E = j\omega\mu\, H, \tag{2.40}$$

$$\nabla \times H = (j\omega\epsilon + \sigma)E, \tag{2.41}$$

with $E = E_o \exp(-\gamma x)\ \hat{y}$ and $H = H_o \exp(-\gamma x)\ \hat{z}$, where $\hat{x}$, $\hat{y}$, $\hat{z}$ are standard unit vectors; we get

$$\gamma E_o = j\omega\mu\, H_o, \tag{2.42}$$

$$\gamma H_o = (j\omega\epsilon + \sigma)E_o, \tag{2.43}$$

giving

$$\gamma = \sqrt{j\omega\mu(j\omega\epsilon + \sigma)} = \alpha + j\beta \tag{2.44}$$

and $\frac{E_0}{H_0} = \eta = \sqrt{\frac{j\omega\mu}{j\omega\epsilon+\sigma}}$. The wave evolves then $E = E_o \exp(-(\alpha + j\beta)x + j\omega t)\ \hat{y}$ and $H = H_o \exp(-(\alpha + j\beta)x + j\omega t)\ \hat{z}$. This is traveling wave in x direction that decays with x with exponent α.

For a typical metal $\sigma = 10^7$ SI units, $\epsilon = \frac{10^{-10}}{4\pi}$ SI units and $\mu = 4\pi \times 10^{-7}$ SI units, and for $\omega = 10^{11}$ SI units, we get $\alpha \sim 10^{-6}$ SI units. Therefore, within a distance of a micron, the wave attenuates to e^{-1} of its value.

Now consider a media interface where region I is a dielectric with permittivity ϵ_1 and region II is a conductor with permittivity ϵ_2 and conductivity σ. For normal incidence, there is transmitted wave in the conductor that decays rapidly inside the conductor as seen above. For E tangential, from previous section, we get $E_{ro} = \frac{1-a}{1+a} E_{io}$. Note $a = \sqrt{\frac{\sigma + j\omega\epsilon_2}{j\omega\epsilon_1}}$. From numbers quoted above $|a| \gg 1$, we get $E_{ro} =\sim E_{io}$ and $E_{to} \sim 0$. In limit $a \to \infty$, we say we have perfect conductor which gives $E_{to} = 0$ and $E_{r0} = -E_{io}$, i.e., from a perfect conductor, we have perfect reflection. The total electric field goes to zero at the interface, i.e., $E_{io} + E_{ro} = 0$ and $E_{to} = 0$ and $H_{io} - H_{ro} = 2H_{io}$ and $H_{to} = 2H_{io}$.

At a rate α, this H_{to} decays to zero such that within a small width $\Delta x \sim \alpha^{-1}$, we drop from $2H_{io}$ to 0. But observe that $\nabla \times H = (j\omega\epsilon_2 + \sigma)E$. This gives $J\Delta x = \sigma E \Delta x = 2H_{io}$. We call the quantity $J_s = J\Delta x$ as the surface current density at the interface, $J_S = J\Delta x = H$, where $H = 2H_{io}$ is the magnetic field at the boundary.

Now consider the case when the angle of incidence is not normal. Then we have two cases:

(a) As in Fig. 2.5, E is tangential and we get $E_{io} = -E_{ro}$, and the normal components of H_{io} and H_{ro} cancel and H at the interface is $2H_{io}\cos\theta$. The surface current density is then $J_S = 2H_{io}\cos\theta$.

(b) As in Fig. 2.6 E is tangential, and we get $E_{io} = E_{ro}$. The tangential components of E_{io} and E_{ro} cancel while $H = 2H_{io}$. The surface current density is then $J_S = 2H_{io}$ along the y direction. Furthermore, normal components of E_{io} and E_{ro} add to give us a net normal field of $2E_{io}\sin\theta$ at the interface. But there is no normal field inside the conductor. If we allude to Fig. 2.4b, this means there is surface charge density of magnitude $\rho_s = 2E_{io}\sin theta\epsilon_2$ at the interface.

In summary, at the conductor interface, the tangential component of E field is zero. The normal component of H field is zero. The tangential component of H field leads to a surface current density and normal component of E field leads to a surface charge density. In both cases, the fields inside the conductor decay rapidly. It is worthwhile writing the fields after reflection in region 1. They arise from interference of incident and reflected waves and take the form

(a)

$$E = -j2E_{io}\exp(-jkz\ \sin\theta)\sin(kx\ \cos\theta)\ \hat{y}, \tag{2.45}$$

$$H = 2H_{io}\exp(-jkz\ \sin\theta)(\cos\theta\cos(kx\ \cos\theta)\ \hat{z} + j\sin\theta\sin(kx\ \cos\theta)\ \hat{x}). \tag{2.46}$$

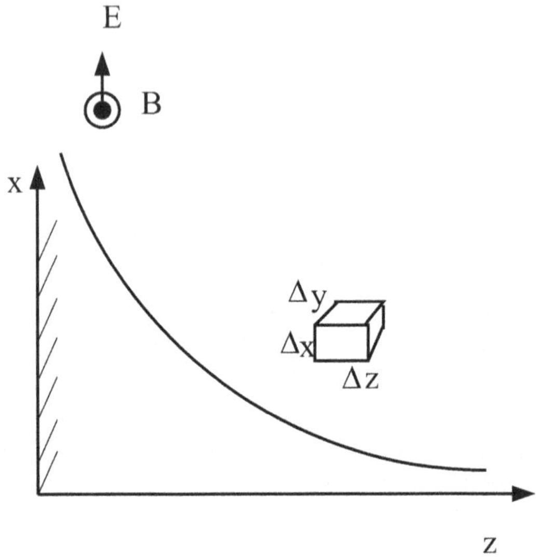

Fig. 2.7 Electric field decaying exponentially inside a conductor

(b)

$$E = 2E_{io}\exp(-jkz\ \sin\theta)(\sin\theta\cos(kx\ \cos\theta)\ \hat{x} \tag{2.47}$$
$$+j\cos\theta\sin(kx\ \cos\theta)\ \hat{z}),$$
$$H = 2H_{io}\exp(-jkz\ \sin\theta)\cos(kx\ \cos\theta)\ \hat{y}. \tag{2.48}$$

2.5 Conductor Losses

Recall from Eq. (2.44) inside a conductor, when $\sigma \gg \omega\epsilon$

$$\gamma = \sqrt{j\omega\mu(j\omega\epsilon + \sigma)} = \alpha + j\beta = \sqrt{\frac{\omega\mu}{\sigma}} + j\sqrt{\frac{\omega\mu}{\sigma}} \tag{2.49}$$

Inside the conductor, electric field is $E(z) = E_0\exp(-\gamma z)$ as shown in Fig. 2.7. Then current density $J(z) = \sigma E(z)$, and integrating it for z from 0 to ∞, we get $J_s = \int J(z)dz = \frac{J(0)}{\gamma} = \frac{\sigma E(0)}{\gamma} = H(0)$. The RMS power loss is a same cubical as shown in Fig. 2.7 is

$$P(z) = \frac{1}{2}|I|^2 R = \frac{1}{2}\underbrace{|J(z)|^2(\Delta z\Delta y)^2}_{|I|^2}\underbrace{\left(\frac{\rho\Delta x}{\Delta z\Delta y}\right)}_{R}. \tag{2.50}$$

Then $\int P(z)dz = \frac{1}{2}J_s^2 R_s$, where $R_s = \sqrt{\frac{\omega\mu_0}{2\sigma}}$ is termed surface resistance. Alternatively, $J_s = J\Delta z$, and root mean square (RMS) power is

$$P = \frac{1}{2}|I|^2 R = \frac{1}{2}\underbrace{(J\Delta z\Delta y)^2}_{|I|^2}\underbrace{\left(\frac{\rho\Delta x}{\Delta z\Delta y}\right)}_{R} = \frac{1}{2}\frac{J_s^2\Delta x\Delta y}{\Delta z\sigma}. \tag{2.51}$$

But note $\frac{1}{\sigma\Delta z} = \sigma^{-1}\alpha = \sqrt{\frac{\omega\mu}{2\sigma}} = R_s$. As an example, if H at the surface of conductor is 2 SI units and ω is $2\pi \times 10^9$, then for $\sigma = 10^7$ SI units, we get $P = 0.56W/m^2$.

2.6 Problems

1. What is electric potential at 1 $A°$ distance from an electron?
2. What is the magnetic field in the center of a circular loop of 1 cm radius carrying 1 ampere of current?
3. If relative permittivity of a medium is 1.4 and incident angle of EM wave is 30°, what is the transmitted angle (with respect to the interface)? (H points inside the plane of reflection)
4. For the above case, find the ratio of amplitude of reflected and incident waves.
5. How far will a 1 GHz EM wave travel in a conductor of conductivity 10^7 SI units?

Chapter 3
Waveguides and Resonators

We have to calculate the eigenfunctions of Maxwell equation, for eigenvalue $\lambda = j\omega$. In the last chapter, we did it for free space and media interface. In this chapter, we look at free space bounded by conductors. This geometry forms waveguides, cavities, and resonators.

3.1 Parallel Plane Waveguides

Consider two parallel conductor planes as shown in Fig. 3.1. Figure **a** shows the EM mode, confined between parallel conductors with E parallel to the conductor plane. Figure **b** shows the EM mode confined between parallel conductors with H parallel to the conductor plane. From the last chapter, after reflection from the interface I, the E and H fields in two cases take the form

(a)

$$E = -j2E_{io} \exp(-jkz \ \sin\theta) \sin(kx \ \cos\theta) \ \hat{y}, \tag{3.1}$$

$$H = 2H_{io} \exp(-jkz \ \sin\theta)(\cos\theta \cos(kx \ \cos\theta) \ \hat{z} \tag{3.2}$$
$$+j \sin\theta \sin(kx \ \cos\theta) \ \hat{x}).$$

(b)

$$E = 2E_{io} \exp(-jkz \ \sin\theta)(\sin\theta \cos(kx \ \cos\theta) \ \hat{x} \tag{3.3}$$
$$+j \cos\theta \sin(kx \ \cos\theta) \ \hat{z}),$$

$$H = 2H_{io} \exp(-jkz \ \sin\theta) \cos(kx \ \cos\theta) \ \hat{y}. \tag{3.4}$$

From Fig. 3.1a, we have, for tangential E and normal H at the interface II to be zero,

N. Khaneja, *Electromagnetic Waves and Optics*,
https://doi.org/10.1007/978-3-031-76144-7_3

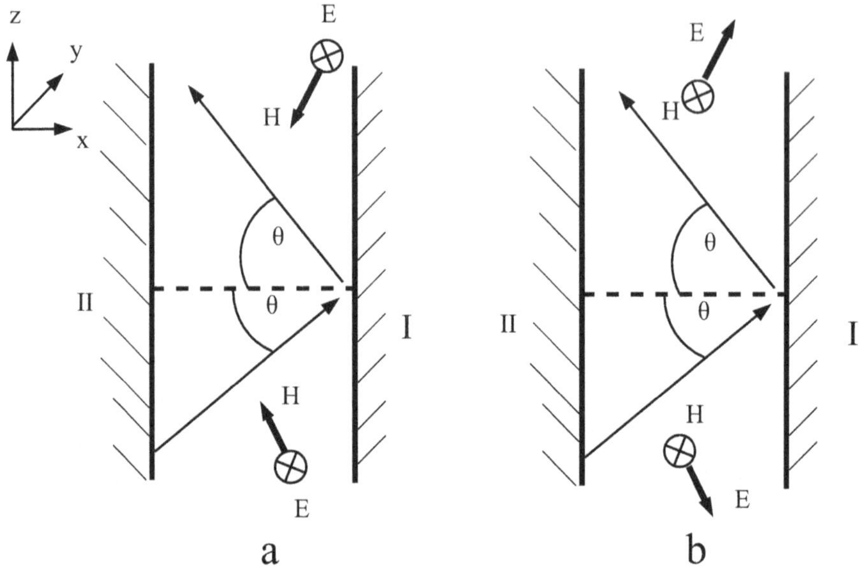

Fig. 3.1 (**a**) EM mode confined between parallel conductors with E parallel to the conductor plane. (**b**) EM mode confined between parallel conductors with H parallel to the conductor plane

$$\cos\theta d = \frac{m\pi}{kd}, \tag{3.5}$$

and $\beta = k\sin\theta$ satisfies; we have $c^2(\beta^2 + (\frac{m\pi}{d})^2) = \omega^2$ from which we get

$$\beta = \sqrt{\left(\frac{\omega}{c}\right)^2 - \left(\frac{m\pi}{d}\right)^2}. \tag{3.6}$$

We have a cutoff frequency when $\beta = 0$. Integer m defines modes and determines θ. Frequency ω defines how high m can be before we reach cutoff.

From Fig. 3.1b, we have, for tangential E and normal H at the interface II to be zero,

$$k\cos\theta d = \frac{m\pi}{kd}, \tag{3.7}$$

and $\beta = k\sin\theta$ satisfies; we have $c^2(\beta^2 + (\frac{m\pi}{d})^2) = \omega^2$ from which we get

$$\beta = \sqrt{\left(\frac{\omega}{c}\right)^2 - \left(\frac{m\pi}{d}\right)^2}. \tag{3.8}$$

We have a cutoff frequency when $\beta = 0$. Integer m defines modes and determine θ. Frequency ω defines how high m can be before we reach cutoff.

3.2 Rectangular Waveguides

We now consider a geometry of a hollow rectangular pipe as shown in Fig. 3.2. It is called a rectangular waveguide. The $x - y$ directions are called transverse, while z direction is called longitudinal. When electric field is all transverse, we call it TE mode and when the magnetic field is all transverse, we call it TM mode.

3.2.1 *TE Mode*

We propose a TE mode propagating along z direction, where we have H_z and E_x, E_y, H_x, H_y, all with a $\exp(-j\beta z)$ dependence, such that using $-\nabla E = j\omega\mu H$ and $\nabla H = j\omega\epsilon E$, we get

$$j\omega\mu H_x = \frac{\partial E_y}{\partial z}, \tag{3.9}$$

$$j\omega\mu H_y = -\frac{\partial E_x}{\partial z}, \tag{3.10}$$

$$j\omega\epsilon E_x = \frac{\partial H_z}{\partial y} - \frac{\partial H_y}{\partial z}, \tag{3.11}$$

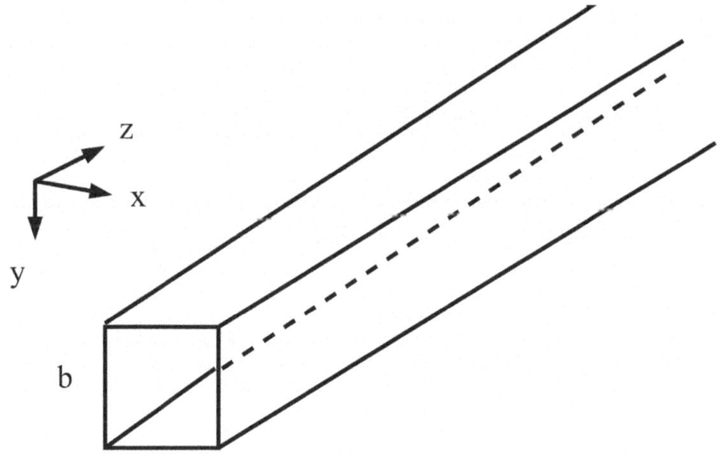

Fig. 3.2 A hollow rectangular pipe called a rectangular waveguide

$$j\omega\epsilon E_y = -\frac{\partial H_z}{\partial x} + \frac{\partial H_x}{\partial z}. \tag{3.12}$$

Substituting Eqs. (3.9) and (3.10) in Eqs. (3.11) and (3.12), we get

$$j\omega\left(\epsilon - \frac{\beta^2}{\omega^2\mu}\right)E_x = \frac{\partial H_z}{\partial y}, \tag{3.13}$$

$$j\omega\left(\epsilon - \frac{\beta^2}{\omega^2\mu}\right)E_y = -\frac{\partial H_z}{\partial x}. \tag{3.14}$$

Substituting the above in Eqs. (3.9) and (3.10), we get

$$\frac{\mu}{\beta}j\omega^2\left(\epsilon - \frac{\beta^2}{\omega^2\mu}\right)H_x = \frac{\partial H_z}{\partial x}, \tag{3.15}$$

$$\frac{\mu}{\beta}j\omega^2\left(\epsilon - \frac{\beta^2}{\omega^2\mu}\right)H_y = \frac{\partial H_z}{\partial y}. \tag{3.16}$$

Observe that the transverse fields E_x, E_y, H_x, H_y all depend on longitudinal component H_z. Observe

$$H_z \propto \cos\left(\frac{m\pi x}{a}\right)\cos\left(\frac{n\pi y}{a}\right), \tag{3.17}$$

$$H_x \propto \sin\left(\frac{m\pi x}{a}\right)\cos\left(\frac{n\pi y}{a}\right), \tag{3.18}$$

$$H_y \propto \cos\left(\frac{m\pi x}{a}\right)\sin\left(\frac{n\pi y}{a}\right), \tag{3.19}$$

$$E_x \propto \cos\left(\frac{m\pi x}{a}\right)\sin\left(\frac{n\pi y}{a}\right), \tag{3.20}$$

$$E_y \propto \sin\left(\frac{m\pi x}{a}\right)\cos\left(\frac{n\pi y}{a}\right). \tag{3.21}$$

That is all, at the boundary, tangential E and normal H are zero; thus, all boundary conditions are satisfied. The mode is written as TE_{mn}.

From $-\nabla E = j\omega\mu H$ and $\nabla H = j\omega\epsilon E$, we get using the fact that $j\omega\mu H_z = \frac{\partial E_x}{\partial y} - \frac{\partial E_y}{\partial z}$, from which we get $\beta = \sqrt{(\frac{\omega}{c})^2 - (\frac{m\pi}{a})^2 - (\frac{n\pi}{b})^2}$. Therefore, mode m, n exists only if $\omega \geq c\sqrt{(\frac{m\pi}{a})^2 + (\frac{n\pi}{b})^2}$.

3.2.2 TM Mode

We propose a TM mode, propagating along z direction, where we have E_z and E_x, E_y, H_x, H_y, all with a $\exp(-j\beta z)$ dependence, such that using $-\nabla E = j\omega\mu H$ and $\nabla H = j\omega\epsilon E$, we get

$$j\omega\mu H_x = \frac{\partial E_y}{\partial z} - \frac{\partial E_z}{\partial y}, \tag{3.22}$$

$$j\omega\mu H_y = -\frac{\partial E_x}{\partial z} + \frac{\partial E_z}{\partial x}, \tag{3.23}$$

$$j\omega\epsilon E_x = -\frac{\partial H_y}{\partial z}, \tag{3.24}$$

$$j\omega\epsilon E_y = \frac{\partial H_x}{\partial z}. \tag{3.25}$$

Substituting Eqs. (3.24) and (3.25) in Eqs. (3.22) and (3.23), we get

$$j\omega(\mu - \frac{\beta^2}{\omega^2\epsilon})H_x = -\frac{\partial E_z}{\partial y}, \tag{3.26}$$

$$j\omega(\mu - \frac{\beta^2}{\omega^2\epsilon})H_y = +\frac{\partial E_z}{\partial x}. \tag{3.27}$$

Substituting the above in Eqs. (3.24) and (3.25), we get

$$\frac{\epsilon}{\beta} j\omega^2\left(\mu - \frac{\beta^2}{\omega^2\epsilon}\right)E_x = \frac{\partial E_z}{\partial x}, \tag{3.28}$$

$$\frac{\epsilon}{\beta} j\omega^2\left(\mu - \frac{\beta^2}{\omega^2\epsilon}\right)E_y = \frac{\partial E_z}{\partial y}. \tag{3.29}$$

Observe that transverse fields E_x, E_y, H_x, H_y all depend on longitudinal component E_z. Observe if

$$H_z \propto \sin\left(\frac{m\pi x}{a}\right) \sin\left(\frac{n\pi y}{a}\right), \tag{3.30}$$

$$H_x \propto \sin\left(\frac{m\pi x}{a}\right) \cos\left(\frac{n\pi y}{a}\right), \tag{3.31}$$

$$H_y \propto \cos\left(\frac{m\pi x}{a}\right) \sin\left(\frac{n\pi y}{a}\right), \tag{3.32}$$

$$E_x \propto \cos\left(\frac{m\pi x}{a}\right) \sin\left(\frac{n\pi y}{a}\right), \tag{3.33}$$

$$E_y \propto \sin\left(\frac{m\pi x}{a}\right)\cos\left(\frac{n\pi y}{a}\right). \tag{3.34}$$

That is all, at the boundary, tangential E and normal H are zero; thus, all boundary conditions are satisfied. The mode is written as TM_{mn}. Again we get $\beta = \sqrt{(\frac{\omega}{c})^2 - (\frac{m\pi}{a})^2 - (\frac{n\pi}{b})^2}$. Therefore mode m, n exists only if $\omega \geq c\sqrt{(\frac{m\pi}{a})^2 + (\frac{n\pi}{b})^2}$.

3.3 Rectangular Cavity

We considered a rectangular waveguide, which is open and indefinite in z direction. If we close its ends and length z length be c. Then the $\exp(-j\beta c)$ wave gets reflected an the end and we have two waves $\propto \exp(-j\beta z)$ and $\propto -\exp(-j\beta z)$, adding the two we get something $\propto \sin(\beta z)$. For electric fields to go to zero at the z ends, we have $\beta = \frac{p\pi}{c}$. The resulting mode is called TE_{mnp} or TM_{mnp} depending on if electric or magnetic field is transverse, i.e., if we have H_z or E_z. This gives mode frequency

$$\omega = \sqrt{\left(\frac{m\pi}{a}\right)^2 + \left(\frac{n\pi}{b}\right)^2 + \left(\frac{p\pi}{c}\right)^2}. \tag{3.35}$$

Writing explicitly the modes for TE_{mnp},

$$H_z \propto \cos\left(\frac{m\pi x}{a}\right)\cos\left(\frac{n\pi y}{a}\right)\sin\left(\frac{p\pi z}{c}\right), \tag{3.36}$$

$$H_x \propto \sin\left(\frac{m\pi x}{a}\right)\cos\left(\frac{n\pi y}{a}\right)\cos\left(\frac{p\pi z}{c}\right), \tag{3.37}$$

$$H_y \propto \cos\left(\frac{m\pi x}{a}\right)\sin\left(\frac{n\pi y}{a}\right)\cos\left(\frac{p\pi z}{c}\right), \tag{3.38}$$

$$E_x \propto \cos\left(\frac{m\pi x}{a}\right)\sin\left(\frac{n\pi y}{a}\right)\sin\left(\frac{p\pi z}{c}\right), \tag{3.39}$$

$$E_y \propto \sin\left(\frac{m\pi x}{a}\right)\cos\left(\frac{n\pi y}{a}\right)\sin\left(\frac{p\pi z}{c}\right). \tag{3.40}$$

Writing explicitly the modes for TM_{mnp},

$$E_z \propto \sin\left(\frac{m\pi x}{a}\right)\sin\left(\frac{n\pi y}{a}\right)\cos\left(\frac{p\pi z}{c}\right), \tag{3.41}$$

$$H_x \propto \sin\left(\frac{m\pi x}{a}\right)\cos\left(\frac{n\pi y}{a}\right)\cos\left(\frac{p\pi z}{c}\right), \tag{3.42}$$

$$H_y \propto \cos\left(\frac{m\pi x}{a}\right)\sin\left(\frac{n\pi y}{a}\right)\cos\left(\frac{p\pi z}{c}\right), \tag{3.43}$$

$$E_x \propto \cos\left(\frac{m\pi x}{a}\right)\sin\left(\frac{n\pi y}{a}\right)\sin\left(\frac{p\pi z}{c}\right), \tag{3.44}$$

$$E_y \propto \sin\left(\frac{m\pi x}{a}\right)\cos\left(\frac{n\pi y}{a}\right)\sin\left(\frac{p\pi z}{c}\right). \tag{3.45}$$

3.4 Cylindrical Cavity

Consider a cylindrical cavity of radius a and height b, where b direction is the longitudinal z direction.

3.4.1 TE Modes

As shown in Fig. 3.3a, we consider TE mode with H_z. In cylindrical coordinates (r, θ, z), we have

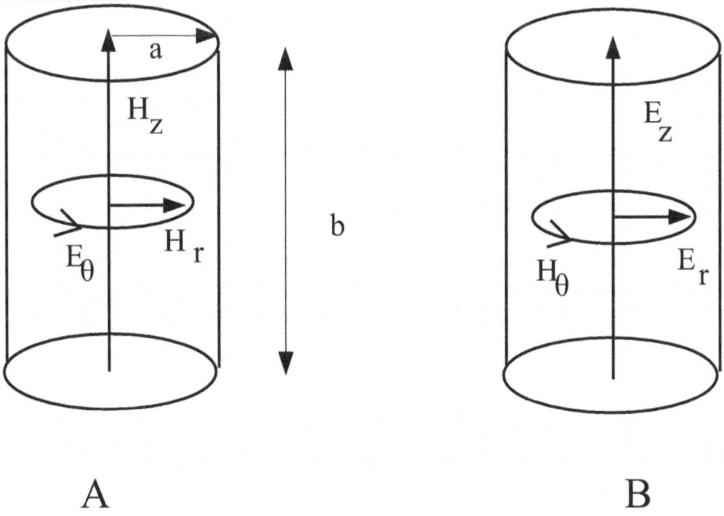

Fig. 3.3 A cylindrical cavity of radius a and height b

$$\frac{\partial}{\partial x} = \cos\theta \frac{\partial}{\partial r} - \frac{\sin\theta}{r}\frac{\partial}{\partial \theta}, \tag{3.46}$$

$$\frac{\partial}{\partial y} = \sin\theta \frac{\partial}{\partial r} + \frac{\cos\theta}{r}\frac{\partial}{\partial \theta}. \tag{3.47}$$

For TE modes from Eqs. (3.13 and 3.14), we can write

$$j\omega\left(\epsilon - \frac{\beta^2}{\mu\omega^2}\right)E_r = \frac{1}{r}\frac{\partial H_z}{\partial \theta}, \tag{3.48}$$

$$j\omega\left(\epsilon - \frac{\beta^2}{\mu\omega^2}\right)E_\theta = -\frac{\partial H_z}{\partial r}. \tag{3.49}$$

from Eqs. (3.15 and 3.16), we get

$$\frac{\mu}{\beta}j\omega^2\left(\epsilon - \frac{\beta^2}{\omega^2\mu}\right)H_r = \frac{\partial H_z}{\partial r}, \tag{3.50}$$

$$\frac{\mu}{\beta}j\omega^2\left(\epsilon - \frac{\beta^2}{\omega^2\mu}\right)H_\theta = \frac{1}{r}\frac{\partial H_z}{\partial \theta}. \tag{3.51}$$

By choosing H_z as a function of r, we get E_θ and H_r that vanish at the boundary. Using the fact that

$$j\omega\mu H_z = \frac{\partial E_x}{\partial y} - \frac{\partial E_y}{\partial z} \propto \frac{\partial^2 H_z}{\partial x^2} + \frac{\partial^2 H_z}{\partial y^2},$$

which gives

From

$$\underbrace{\left(\frac{\omega^2}{c^2} - \beta^2\right)}_{l^2} H_z = \left(\frac{\partial^2}{\partial r^2} + \frac{1}{r}\frac{\partial}{\partial r}\right)H_z.$$

The solution is $J_0(lr)$. If Z_1, Z_2, Z_n are zeros of $J_0'(x)$ (other than trivial zeros at 0), then modes are defined as $la = Z_i$. This ensures that E_θ and H_r vanish at the boundary.

This defines $\beta^2 = \sqrt{(\frac{\omega^2}{c^2} - (\frac{Z_i}{a})^2}$. We have a wave along z with dependence $\exp(-j\beta x)$. Superimposing on it a reflected wave $\exp(-j\beta x)$, we get an electric field that has z dependence $\sin(\beta z)$ and magnetic field $\cos(\beta z)$. By choosing $\beta = \frac{p\pi}{b}$, we get a standing wave mode indexed by p and Z_i with

$$\omega = c\sqrt{\left(\frac{p\pi}{b}\right)^2 + \left(\frac{Z_i}{a}\right)^2}, \tag{3.52}$$

The resulting fields are shown in Fig. 3.3a.

We choose H_z as a function of r only. This is not the only solution. We can choose

$$H_z = f(r)\cos n\theta. \tag{3.53}$$

Then we get

$$-\underbrace{\left(\frac{\omega^2}{c^2} - \beta^2\right)}_{l^2} H_z = \left(\frac{\partial^2}{\partial r^2} + \frac{1}{r}\frac{\partial}{\partial r} + \frac{1}{r^2}\frac{\partial^2}{\partial \theta^2}\right) H_z.$$

$$\left(-l^2 + \frac{n^2}{r^2}\right) f(r) = \left(\frac{\partial^2}{\partial r^2} + \frac{1}{r}\frac{\partial}{\partial r}\right) f(r).$$

The solution is $J_n(lr)$. If $Z_1, Z_2, \ldots, Z_n$ are zeros of $J_n'(x)$ (other than trivial zeros at 0), then modes are defined as $la = Z_i$. This ensures that E_θ and H_r vanish at the boundary.

This defines $\beta^2 = \sqrt{(\frac{\omega^2}{c^2} - (\frac{Z_i}{a})^2}$. We have a wave along z with dependence $\exp(-j\beta z)$. Superimposing on it a reflected wave $\exp(-j\beta z)$, we get an electric field that has z dependence $\sin(\beta z)$ and radial magnetic field $\cos(\beta z)$. By choosing $\beta = \frac{p\pi}{b}$, we get a standing wave mode indexed by p and Z_i with

$$\omega = c\sqrt{\left(\frac{p\pi}{b}\right)^2 + \left(\frac{Z_i}{a}\right)^2}. \tag{3.54}$$

The resulting fields are shown in Fig. 3.3a.

3.4.2 TM Modes

As shown in Fig. 3.3b we consider TM mode with E_z. For TM modes from Eqs. (3.28 and 3.29), we can write

$$j\omega\left(\mu - \frac{\beta^2}{\epsilon\omega^2}\right) H_r = \frac{1}{r}\frac{\partial E_z}{\partial \theta}, \tag{3.55}$$

$$j\omega\left(\mu - \frac{\beta^2}{\epsilon\omega^2}\right) H_\theta = \frac{\partial E_z}{\partial r}. \tag{3.56}$$

From Eqs. (3.26 and 3.27), we get

$$\frac{\epsilon}{\beta} j\omega^2 \left(\mu - \frac{\beta^2}{\omega^2 \epsilon}\right) E_r = \frac{\partial E_z}{\partial r}, \tag{3.57}$$

$$\frac{\epsilon}{\beta} j\omega^2 \left(\mu - \frac{\beta^2}{\omega^2 \epsilon}\right) E_\theta = \frac{1}{r} \frac{\partial E_z}{\partial \theta}. \tag{3.58}$$

By choosing E_z as a function of r, we get E_θ and H_r that vanish at the boundary.

$$j\omega\epsilon E_z = \frac{\partial E_y}{\partial x} - \frac{\partial E_x}{\partial y} \propto \frac{\partial^2 E_z}{\partial x^2} + \frac{\partial^2 E_z}{\partial y^2},$$

which gives

From

$$\underbrace{\left(\frac{\omega^2}{c^2} - \beta^2\right)}_{l^2} E_z = \left(\frac{\partial^2}{\partial r^2} + \frac{1}{r}\frac{\partial}{\partial r}\right) H_z.$$

The solution is $J_0(lr)$. If Y_1, Y_2, Y_n are zeros of $J_0(x)$, then modes are defined as $la = Y_i$. This ensures that E_z vanishes at the boundary.

This defines $\beta^2 = \sqrt{(\frac{\omega^2}{c^2} - (\frac{Y_i}{a})^2}$. We have a wave along z with dependence $\exp(-j\beta x)$. Superimposing on it a reflected wave $\exp(-j\beta z)$, we get a radial electric field has z dependence $\sin(\beta z)$, z electric field has dependence $\cos(\beta z)$, and magnetic field $\cos(\beta z)$. By choosing $\beta = \frac{p\pi}{b}$, we get a standing wave mode indexed by p and Y_i with

$$\omega = c\sqrt{\left(\frac{p\pi}{b}\right)^2 + \left(\frac{Y_i}{a}\right)^2}. \tag{3.59}$$

The resulting fields are shown in Fig. 3.3b. We choose E_z as function of r only. This is not the only solution.

We choose

$$E_z = f(r)\cos n\theta. \tag{3.60}$$

$$j\omega\epsilon E_z = \frac{\partial E_y}{\partial x} - \frac{\partial E_x}{\partial y} \propto \frac{\partial^2 E_z}{\partial x^2} + \frac{\partial^2 E_z}{\partial y^2},$$

which gives From

$$\underbrace{\left(\frac{\omega^2}{c^2} - \beta^2\right)}_{l^2} E_z = \left(\frac{\partial^2}{\partial r^2} + \frac{1}{r}\frac{\partial}{\partial r} + + \frac{1}{r^2}\frac{\partial^2}{\partial \theta^2}\right) E_z.$$

$$\left(-l^2 + \frac{n^2}{r^2}\right) f(r) = \left(\frac{\partial^2}{\partial r^2} + \frac{1}{r}\frac{\partial}{\partial r}\right) f(r).$$

The solution in the annulus is $J_n(lr)$. If $Y_1, Y_2, \ldots, Y_n$ are zeros of $J_n(x)$ (other than trivial zeros at 0), then modes are defined as $la = Y_i$. This ensures that E_θ and H_r vanish at the boundary.

This defines $\beta^2 = \sqrt{(\frac{\omega^2}{c^2} - (\frac{Y_i}{a})^2}$. We have a wave along z with dependence $\exp(-j\beta z)$. Superimposing on it a reflected wave $\exp(-j\beta z)$, we get a transverse electric field that has z dependence $\sin(\beta z)$ and transverse magnetic field $\cos(\beta z)$. By choosing $\beta = \frac{p\pi}{b}$, we get a standing wave mode indexed by p and Y_i with

$$\omega = c\sqrt{\left(\frac{p\pi}{b}\right)^2 + \left(\frac{Y_i}{a}\right)^2}. \tag{3.61}$$

3.5 Quality Factor of Resonator

We discussed cavities and their modes, say mode ϕ with eigenvalue $\lambda = j\omega$. The mode will just evolve as $\exp(j\omega t)\phi(r)$. It means the mode will live in the cavity forever. That is not true. The mode corresponds to J on the walls which means resistive losses. This means we really have $\lambda = -\alpha + j\omega$, where λ is a decay constant. How do we calculate α? We can calculate how much power P is dissipated on the walls using the formula P per unit area as $\frac{1}{2}J_s^2 R_s$, and we know that the energy E of the mode decreases by $\Delta E = P\Delta t$ in unit time Δt. If A is the amplitude of mode, then in time Δt, we have

$$\frac{\Delta A}{A} = \alpha \Delta t. \tag{3.62}$$

Energy $E \propto A^2$ and change of energy $\Delta E = P\Delta t \propto 2A\Delta A$, which gives $2\alpha = \frac{P}{E}$. We define the quality factor Q of the cavity as

$$Q = \frac{\alpha}{\omega}. \tag{3.63}$$

3.6 Problems

1. For a waveguide of cross section 1 cm^2, find the β for the lowest TM mode at 30 GHz frequency.
2. For a waveguide of cross section 1 cm^2, find the β for the lowest TE mode at 30 GHz frequency.
3. Find the resonance frequency of TE_{112} mode in a rectangular cavity 1 cm^3.
4. Find the resonance frequency of TM_{111} mode in a rectangular cavity 1 cm^3.
5. In a cylindrical cavity of 1 cm radius and height, find the lowest resonant frequency of TE and TM modes.

Chapter 4
Antennas

4.1 Vector Potential

In this chapter, we study Maxwell equations, when they are driven by a current source also called an antenna [2, 6, 8, 17, 28]. Recall that in the presence of current source, the Maxwell equations take the form

$$\mu \frac{\partial H}{\partial t} = -\nabla \times E, \tag{4.1}$$

$$\epsilon \frac{\partial E}{\partial t} = \nabla \times H - J(r, t), \tag{4.2}$$

where each spatial location r indices a vector entry in the linear system as in Eq. (2.16). The vector entries are E and B fields. The goal is to calculate $E(r, t)$ and $B(r, t)$ resulting from an current input $J(r, t)$. That is, solve a linear system (2.16) for an input $u(t)$. For instance, an input of the type $u(t) = \exp(j\omega t)$. Earlier, we showed that one way to approach this is to find the modes of the system and the corresponding eigenvalues. That is, we have to solve for $Ax = \lambda x$ in Eq. (2.17), i.e., solve for

$$-\frac{1}{\mu}\nabla \times E = \lambda H, \tag{4.3}$$

$$\frac{1}{\epsilon}\nabla \times H = \lambda E. \tag{4.4}$$

In this section, we adopt a more direct approach to solve for the Maxwell equations given antenna excitation. The antenna we consider is pint current source, with current density J_0 located at origin with current in z direction, as shown in Fig. 4.1. It has x-y area A_e and z length dl such that $A_e J = i_0$. We first observe that since $\nabla \cdot B = 0$, means it can be written as $B = \nabla \times \mathbf{A}$. From (4.1), we get

N. Khaneja, *Electromagnetic Waves and Optics*,
https://doi.org/10.1007/978-3-031-76144-7_4

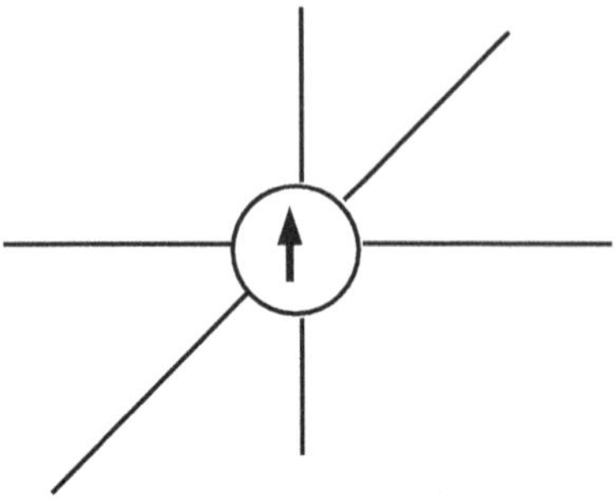

Fig. 4.1 Point current source called Hertz dipole

$$\nabla \times \left(\frac{\partial \mathbf{A}}{\partial t} + E \right) = 0, \tag{4.5}$$

which implies that

$$\frac{\partial \mathbf{A}}{\partial t} + E = -\nabla V, \tag{4.6}$$

$$E = -\frac{\partial \mathbf{A}}{\partial t} - \nabla V. \tag{4.7}$$

We still have some independence in the choice of V and $\mathbf{A}$ that result in the same E and B $V \rightarrow V - \frac{\partial \phi}{\partial t}$ and $\mathbf{A} \rightarrow \mathbf{A} + \nabla \phi$ does not change E and B. We call these gauge degrees of freedom. We use these degrees of freedom to choose $\mathbf{A}$, V such that

$$\frac{\partial V}{\partial t} + \frac{\partial \mathbf{A}_x}{\partial x} + \frac{\partial \mathbf{A}_y}{\partial y} + \frac{\partial \mathbf{A}_z}{\partial z} = 0. \tag{4.8}$$

This is called a Lorentz gauge. This gives from (4.2) that for $i \in \{x, y, z\}$.

$$\nabla^2 \mathbf{A}_i - \frac{\partial^2 \mathbf{A}_i}{c^2 \partial t^2} = \mu_0 J_i(x, t). \tag{4.9}$$

Recall that we assumed current density is in the z direction, and we take a point current source stationed at the origin such that $J_z(x)$ is constant at $J(t)$ over an infinitesimal volume; we take it as a $J(t) = \exp(j\omega t) J_0$, as shown in Fig. 4.2a. The current source is also called a **Hertz dipole**.

$$\frac{\partial^2 \mathbf{A}_z}{c^2 \partial t^2} - \nabla^2 \mathbf{A}_z - = \mu_0 J_z(x, t). \tag{4.10}$$

We first solve the above equations when the current source is stationary and solve for

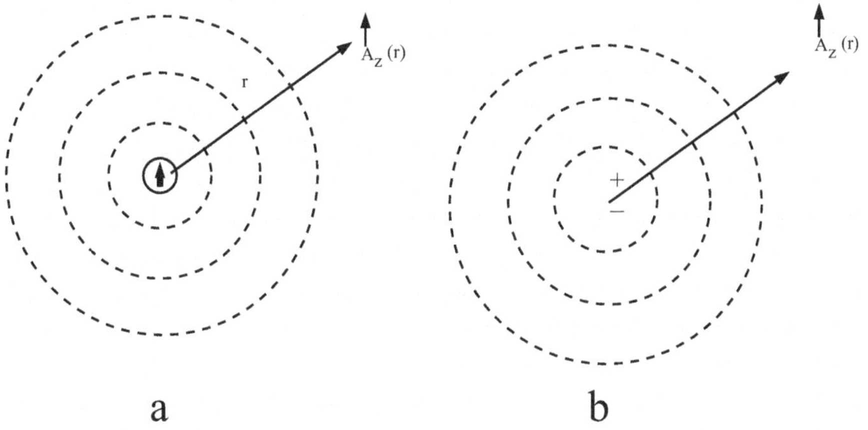

Fig. 4.2 (**a**) and (**b**)Vector potential **A** due to Hertz dipole

$$-\nabla^2 \mathbf{A}_z = \mu_0 J_z(x). \tag{4.11}$$

We can see that $\mathbf{A}_z(r) = \frac{C}{r}$ satisfies the equation $\nabla^2 \mathbf{A}_z = 0$. How do we determine C, we get $\nabla^2 \mathbf{A}_z = \nabla \cdot (\nabla \mathbf{A}_z)$, where $\nabla \mathbf{A}_z = -\frac{C\hat{r}}{r^2}$. Using Stokes' theorem, we get $-\nabla \mathbf{A}_z(4\pi r^2) = \mu_0 J_0 V_0$, which gives $C = \frac{\mu_0 J_0 V_0}{4\pi}$, and we get

$$\mathbf{A}_z(r) = \frac{\mu_0 J_0 V_0}{4\pi r}\,\hat{\mathbf{z}}. \tag{4.12}$$

Now we claim that in the presence of time-varying $J(t)$, we have

$$\mathbf{A}_z(r, t) = \frac{\mu_0 J(t - \frac{r}{c}) V_0}{4\pi r}\,\hat{\mathbf{z}}. \tag{4.13}$$

It is just verified that Eq. (4.13) satisfies (4.10). Writing $J_0 V_0 = i dl$ where i is the current and dl the length of the current element, we can write [11]

$$\mathbf{A}_z(r, t) = \frac{\mu_0\, i(t - \frac{r}{c}) dl}{4\pi r}\,\hat{\mathbf{z}}. \tag{4.14}$$

We can calculate B by using $B = \nabla \times \mathbf{A}$ and $i(t - \frac{r}{c}) = i_0 \exp(j\omega(t - \frac{r}{c})) = i_0 \exp(j\omega(t - kr))$, where $k = \frac{\omega}{c}$.

$$B(r, t) = \frac{\mu_0 dl}{4\pi}\left(\frac{i(t - \frac{r}{c})}{r^2} + \frac{i'(t - \frac{r}{c})}{cr}\right)\hat{\boldsymbol{\phi}}, \tag{4.15}$$

$$B(r,t) = \exp(j\omega t) \underbrace{\frac{\mu_0 i_0 dl \exp(-jkr) \sin\theta}{4\pi r} \left(\frac{1}{r} + jk\right) \hat{\boldsymbol{\phi}}}_{B(r)}. \tag{4.16}$$

Thus, we get the steady-state value $B(r)$. To get the steady-state value $E(r)$, observe

$$j\omega\mu H(r) = -\nabla \times E(r), \tag{4.17}$$

$$j\omega\epsilon E(r) = \nabla \times H(r) - J(r). \tag{4.18}$$

From (4.18), we have for r away from origin

$$\frac{j}{c^2}\omega E = \nabla \times B, \tag{4.19}$$

which gives

$$E(r,t)$$
$$= \exp(j\omega t) \underbrace{\frac{c^2 \mu_0 i_0 dl \exp(-jkr)}{4\pi r\omega} \{ \left(-jk^2 + \frac{3k}{r} - \frac{3j}{r^2}\right) \sin\theta\, \hat{\boldsymbol{\theta}} + \left(\frac{2k}{r} - \frac{2j}{r^2}\right) \hat{\mathbf{z}} \}}_{E(r)}. \tag{4.20}$$

At large distances $r \gg \lambda$, we only keep terms of order r^{-1} giving

$$B(r) = \frac{j\, k\, \mu_0\, i_0\, dl\ \exp(-jkr)\ \sin\theta}{4\pi r} \hat{\boldsymbol{\phi}}, \tag{4.21}$$

$$E(r) = \frac{-j\, k^2\, i_0\, dl\ \exp(-jkr)\ \sin\theta}{4\pi r\omega\epsilon} \hat{\boldsymbol{\theta}}. \tag{4.22}$$

4.1.1 Dipole Radiation

We considered a current source and radiation due to that. Now consider an oscillating dipole, where two opposite charges with charge q go back and forth along say z axis with a maximum separation of d. The dipole moment

$$D(t) = qd\cos(\omega t). \tag{4.23}$$

This is as shown in Fig. 4.2b, where $\frac{dD}{dt}$ constitutes a current. For simplicity, take

$$D(t) = qd \exp(j\omega t). \tag{4.24}$$

Then everything we did for point current source carries over and we make the correspondence $i_0\, dl = jqd\omega$. Then in the expression for $\mathbf{A}_z(r, t)$, $B(r, t)$, $E(r, t)$ in Eqs. (4.14, 4.21, 4.22), we just substitute

$$i_0\, dl = jqd \;\; \omega. \tag{4.25}$$

This gives us radiation due to an oscillating dipole.

4.2 Radiated Power and Antenna Resistance

Consider an enclosed volume V as shown in Fig. 4.3. The volume contains electric and magnetic fields, and the enclosed energy is

$$\mathcal{E} = \frac{1}{2}\int_V \epsilon_0 |E|^2 + \mu_0 |H|^2. \tag{4.26}$$

Then the change of energy with time

Fig. 4.3 Power outflux from an enclosed volume V and surface S

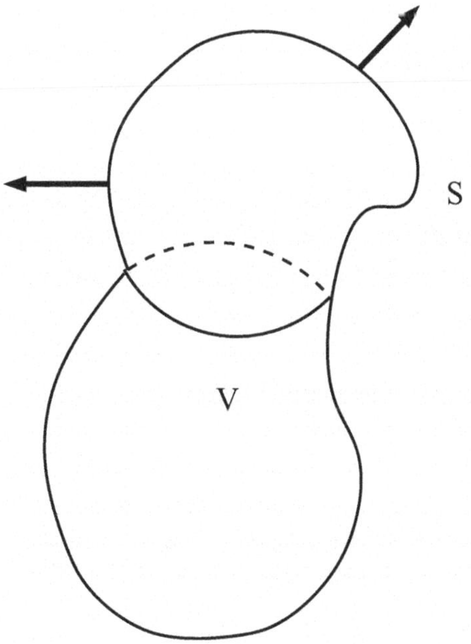

$$\frac{\partial \mathcal{E}}{\partial t} = \int_V Re\left(\epsilon_0 \frac{\partial E}{\partial t} \cdot E^* + \mu_0 \frac{\partial H}{\partial t} \cdot H^*\right), \tag{4.27}$$

$$= \int_V Re(\nabla \times H \cdot E^* - \nabla \times E \cdot H^*), \tag{4.28}$$

$$= \frac{1}{2} \int_V Re(\nabla \times (H^* \times E)), \tag{4.29}$$

$$= \frac{1}{2} \int_S Re(H^* \times E).dA, \tag{4.30}$$

where we use Maxwell equations to substitute for $\frac{\partial E}{\partial t}, \frac{\partial H}{\partial t}$. We use the identity $a \cdot (b \times c) = \frac{1}{2}(c \cdot (a \times b) - b \cdot (a \times c))$. The last equation uses Stokes' theorem. The quantity

$$P = \frac{1}{2} Re(E \times H^*), \tag{4.31}$$

denotes power leaving the surface per unit area. It is also called Poynting vector.

If we integrate P for E and B as in Eqs. (4.21 and 4.22) for a Hertz dipole over a large sphere, we get

$$P = \frac{1}{2} \frac{(ki_0 dl)^2}{6\pi c\epsilon} = i_0^2 R. \tag{4.32}$$

where

$$R = \frac{(kdl)^2}{6\pi c\epsilon}. \tag{4.33}$$

If $dl = \lambda/2$, then $R \sim \sqrt{\frac{\mu_o}{\epsilon_0}} \sim 50$ SI units. It is called radiation resistance of antenna. The power loss due to radiation is seen as a resistance by antenna circuit.

4.3 Antenna Arrays

Now consider an array of Hertz dipoles as shown in Fig. 4.4. Let the total be $N + 1$. Let d be the separation between them. Then $\mathbf{A}_z(r, t)$, $B(r, t)$, $E(r, t)$ due to each dipole carries a factor $\propto \exp(ikr)$, where r between successive dipoles differs by $\Delta r = d \cos\phi$, as shown in Fig. 4.4. This phase factor between successive dipoles differs by a factor $Z = \exp(ik\theta)$, where $\theta = d\cos\phi = d\sin\upsilon$, where $\upsilon = \frac{\pi}{2} - \phi$. Then adding all the phase factors for a point far than the size of array, we get $p = kd\cos\phi$,

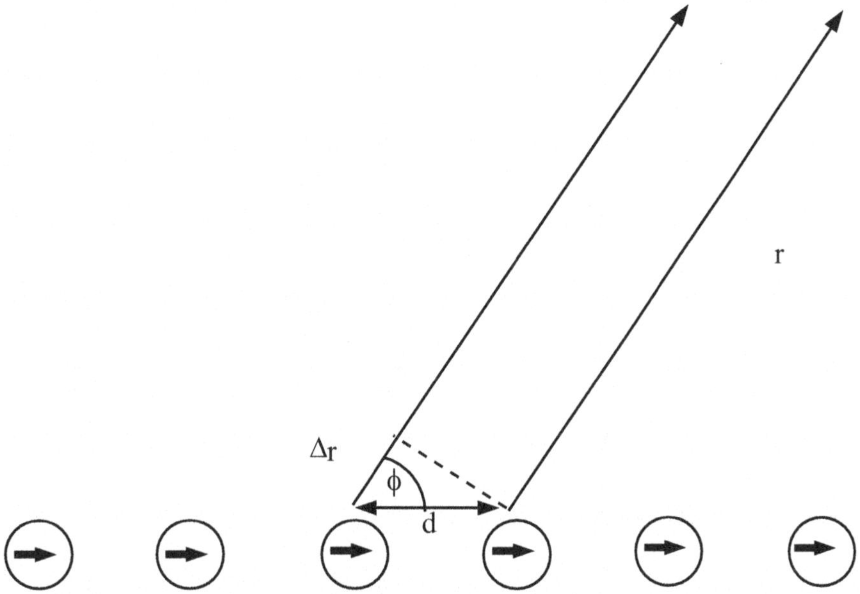

Fig. 4.4 1D array of Hertz dipoles with a separation d between them

$$\eta = \sum_{k=0}^{N} Z^k = \exp\left(j\frac{(N-1)p}{2}\right)\frac{\sin\frac{Np}{2}}{\sin\frac{p}{2}}. \tag{4.34}$$

The factor $f(p) = \frac{\sin\frac{Np}{2}}{\sin\frac{p}{2}}$ gives the dependency of $|\eta|$ as a function of p. It peaks at $p = 0$, where its value is N. Then

$$\frac{|\eta|}{N} = |\frac{\sin\frac{Np}{2}}{N\sin\frac{p}{2}}| \sim |\,\text{sinc}\left(\frac{Np}{2}\right)|, \tag{4.35}$$

where $|\,\text{sinc}(p)|$ is as shown in Fig. 4.5. Few lobes down from origin, the function significantly diminishes. Let us say the array is 1000 wavelengths long which means $Np = 1000\cos\phi$. If we are ten lobes down means $\frac{Np}{2} = 10\pi$ or $\cos\phi = \frac{\pi}{50}$ or $\sin\upsilon = \frac{\pi}{50}$ a small angle. The antenna array fires vertically up as shown in A in Fig. 4.6.

Now suppose we introduce a phase difference in the current in the successive elements of array the form $\exp(i\delta)$, then writing $\delta = -kd\sin\upsilon_0$, we get $p = kd(\sin\upsilon - \sin\upsilon_0)$. Now $p = 0$ when $\upsilon = \upsilon_0$. Therefore, this array will not fire vertically but rather at an angle υ_0. When $\upsilon_0 = \frac{\pi}{2}$, we get an array that fires at the end called an end-fire array as shown in B in Fig. 4.6.

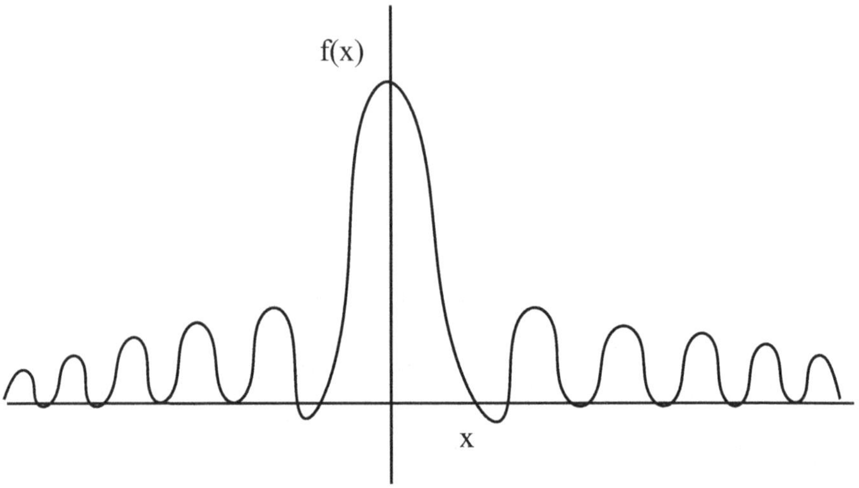

Fig. 4.5 Figure depicting the function $f(x) = |\operatorname{sinc}(x)|$

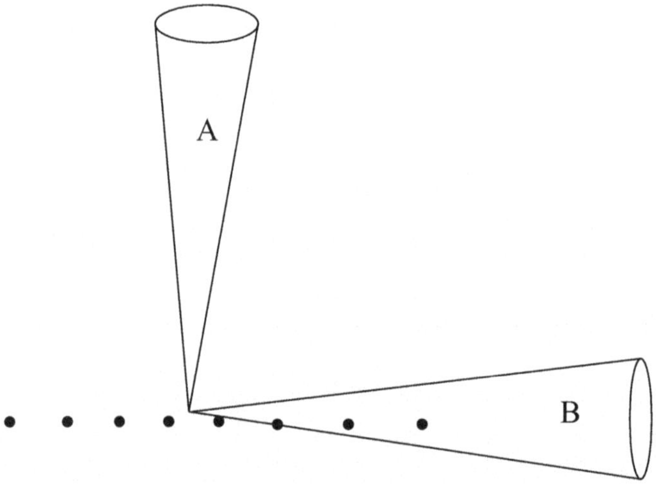

Fig. 4.6 Antenna array firing vertically *A* or horizontally *B* depending on phase difference of the current in array elements

4.3.1 2D arrays

Figure 4.7a shows a 2D array of Hertz dipoles. Figure 4.7b shows the array laid on $x - y$ plane and beam direction making angle θ with z axis and ϕ with x axis. Let d be the spacing between dipoles. Let

$$p_1 = d \sin\theta \cos\phi, \tag{4.36}$$

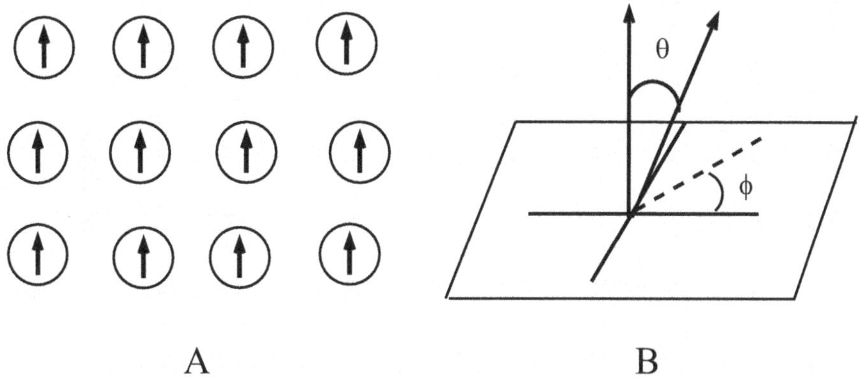

Fig. 4.7 (**a**) 2D array of Hertz dipoles. (**b**) Array laid on $x - y$ plane and beam direction making angle θ with z axis and ϕ with x axis

$$p_2 = d \sin\theta \sin\phi. \tag{4.37}$$

$Z_1 = \exp(jp_1)$ and $Z_2 = \exp(jp_2)$. Then adding phases over antennas, we get

$$\eta = \sum_{k,l} Z_1^k Z_2^l = \exp\left(j\frac{(N-1)p_1 + (N-1)p_2}{2}\right)\frac{\sin\frac{Np_1}{2}}{\sin\frac{p_1}{2}}\frac{\sin\frac{Np_2}{2}}{\sin\frac{p_2}{2}}. \tag{4.38}$$

Let us say the array is 1000×1000 wavelengths in area which means $Np_1 = 1000 \sin\theta\cos\phi$ and $Np_2 = 1000 \sin\theta \sin\phi$. If we have $\sin\theta = \frac{\pi}{50}$, then one of the $\frac{Np_1}{2}$ or $\frac{Np_1}{2}$ is $\sim 10\pi$ or ten lobes down, and η diminishes significantly. Therefore, all intensity is concentrated in θ very small, antenna array fires vertically up.

4.3.2 Antennas and Finite Interfaces

In previous chapters, we studied how EM waves reflect of conducting interface. For ease of calculation, we took the conductor plane to be infinite. How about finite conductor interfaces as in mirrors? To see how mirror reflects EM waves say light (optical frequency 10^{15} Mhz), which is sub-micrometer wavelength, imagine an infinite plane. Incident light induces current density J_S on the plane, which acts as an antenna array. If we partition the plane into say rectangles of say 1000×1000 wavelengths in area (this means mm^2 subblocks for light), then these rectangles will fire vertically as discussed in previous section and shown in Fig. 4.8a. Then suppose the real interface we are given is say region II in Fig. 4.8a. Then if we remove remaining regions, one by one, we find that since their respective antenna did not interfere (at moderate distances up to meter) with the radiation of region II, the

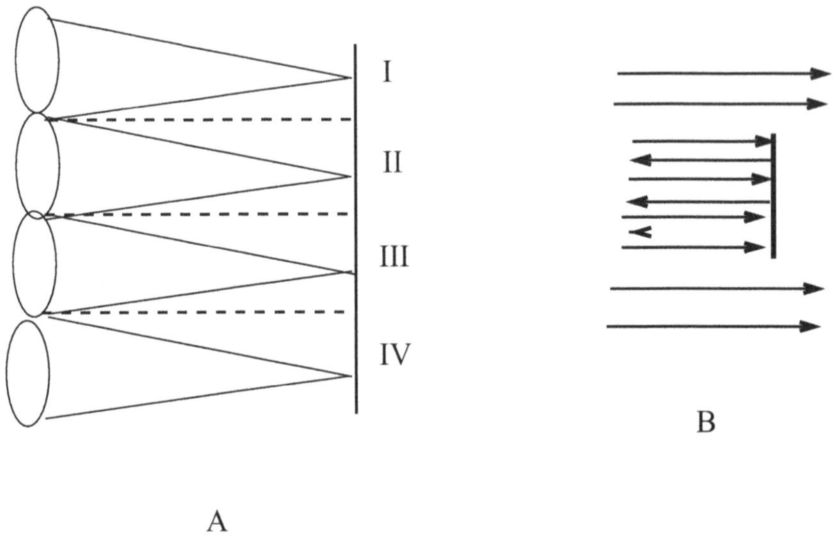

Fig. 4.8 (**a**) Reflection of a conducting plane as finite size antenna arrays firing vertically. (**b**) Ray diagram for incident and reflected light

radiation in the vertical line of sight of region II is uninterrupted. All we are saying is that pretend the plane is infinite, calculate the reflected wave, and only take the reflected wave in vertical line of sight of region II. That is, to say, we can draw ray diagrams as Fig. 4.8b, which is the starting point of geometric optics. Instead of light if we had radio waves, then a mm^2 size reflector will act as a point antenna and send waves in all directions. Then we cannot follow the rules of geometric optics.

4.4 Problems

1. Show that $\nabla \times (\nabla \times E) = -\nabla^2 E + \nabla(\nabla \cdot E)$.
2. Show that $\nabla \cdot (A \times B) = A \cdot (\nabla \times B) - B \cdot (\nabla \times A)$.
3. Write Maxwell equations in terms of vector potential. Use Lorentz gauge.
4. Find E and B for 10 GHz current dipole (at origin) at the point $(r, 0, 0)$, where $r = 1$ km, for 1 watt of radiated power.
5. For an antenna array of 1000 wavelength length, find the beamwidth.

Chapter 5
Transmission Lines

5.1 Coaxial Transmission Lines

In the previous chapter, we discussed a special kind of waveguide, a hollow pipe. In this chapter, we describe a hollow pipe containing a conductor inside it as shown in Fig. 5.1a. This is called a coaxial cable [22, 25] or just coax. Such a structure supports a novel EM mode called the transverse electric and magnetic (TEM) mode, which was absent in a hollow pipe. The electric field is radial and magnetic field is circular as shown in Fig. 5.1b, which shows a cross section of coax.

The radial electric field produces an electric charge Q of opposite polarity in the inner and outer conductors such that the resulting voltage $V(x) = E(x)a$ can be related to the charge Q. Then we can relate V and Q as $V(x)C\Delta x = Q(x)$, where C is the capacitance per unit length.

Furthermore, $(E(x + \Delta x) - E(x))a = -\frac{\partial B}{\partial t} A_r$, where $A_r = a\Delta x$ is the area. But $BA_r = L\Delta x I$, where L is the inductance per unit length and $I(x)$ is the current in the inner conductor. This is written as

$$V(x + \Delta x) - V(x) = L\Delta x \frac{dI}{dt}, \tag{5.1}$$

$$-\frac{\partial V}{\partial x} = L\frac{dI(x)}{dt}. \tag{5.2}$$

Similarly,

$$I(x) - I(x + \Delta x) = \frac{dQ(x)}{dt}, \tag{5.3}$$

$$\frac{dQ(x)}{dt} = C\Delta x \frac{dV(x)}{dt}, \tag{5.4}$$

N. Khaneja, *Electromagnetic Waves and Optics*,
https://doi.org/10.1007/978-3-031-76144-7_5

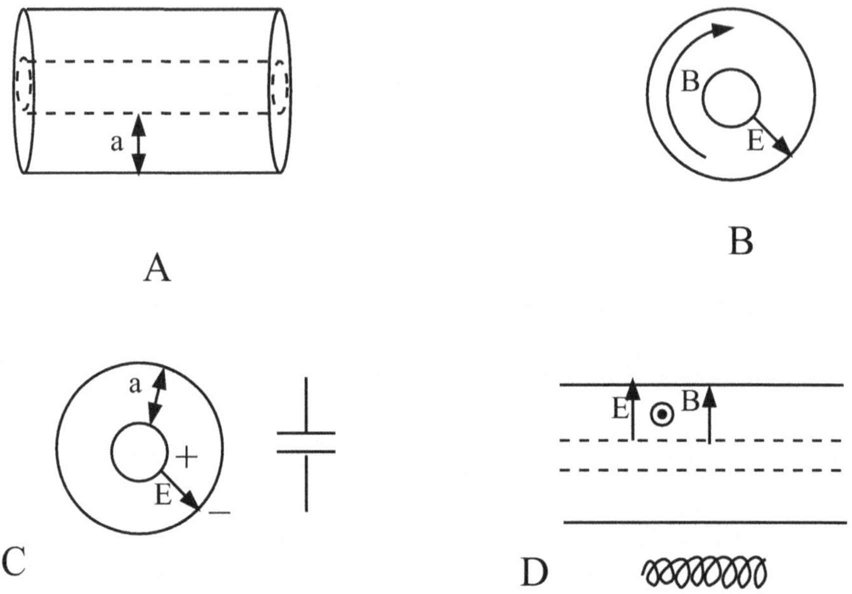

Fig. 5.1 (**a**) Coaxial cable. (**b**) Cross section with radial E and circular B fields. (**c**) How radial E gives capacitance. (**d**) How circular B gives inductance

$$-\frac{\partial I}{\partial x} = C\frac{dV(x)}{dt}. \tag{5.5}$$

Thus, we get two equations of the transmission line

$$-\frac{\partial V}{\partial x} = L\frac{dI(x)}{dt}, \tag{5.6}$$

$$-\frac{\partial I}{\partial x} = C\frac{dV(x)}{dt}. \tag{5.7}$$

Now we drive the transmission line with a voltage source $V = V_0 \exp(j\omega t)$, see Fig. 5.2b. Then that gives steady-state values $V(x)\exp(j\omega t)$ and $I(x)\exp(j\omega t)$ which satisfy

$$-\frac{\partial V}{\partial x} = j\omega L I(x), \tag{5.8}$$

$$-\frac{\partial I}{\partial x} = j\omega C V(x). \tag{5.9}$$

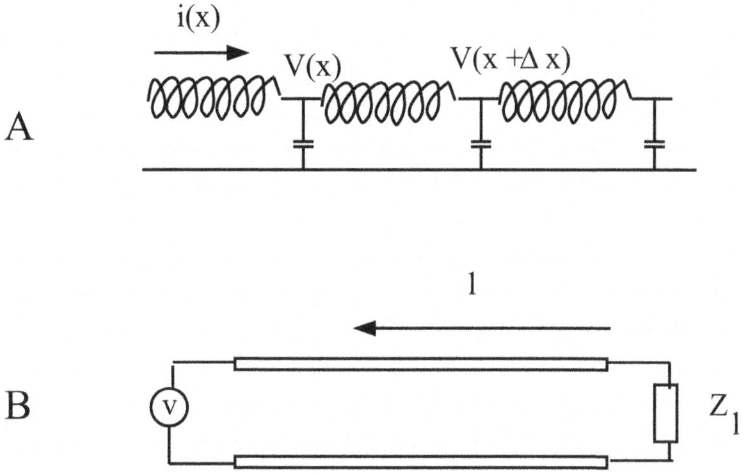

Fig. 5.2 (**a**) Electrical equivalent of a transmission line with distributed inductors and capacitors. (**b**) Voltage source at one end of the line (called generator end) and load at other end

See Fig. 5.2b, where we take load at right and voltage generator at left. Let us take x going from right to left. Then the above equations read

$$\frac{\partial V}{\partial x} = j\omega L I(x), \tag{5.10}$$

$$\frac{\partial I}{\partial x} = j\omega C V(x). \tag{5.11}$$

$$\frac{\partial^2 V}{\partial x^2} = -\omega^2 LCV(x) = -\gamma^2 V(x), \tag{5.12}$$

where $\gamma = j\beta$ where $\beta = \omega\sqrt{LC}$ and let $Z_0 = \sqrt{\frac{L}{C}}$.

$$V(x) = V_+ \exp(j\beta x) + V_- \exp(-j\beta x), \tag{5.13}$$

$$I(x) = \frac{1}{Z_0}(V_+ \exp(j\beta x) - V_- \exp(-j\beta x)) \tag{5.14}$$

$$= \frac{1}{Z_0}(I_+ \exp(j\beta x) - I_- \exp(-j\beta x)). \tag{5.15}$$

5.2 Impedance

Then for $\Gamma = \frac{V_-}{V_+}$,

$$Z(0) = Z_L = \frac{V(0)}{I(0)} = Z_0 \frac{1+\Gamma}{1-\Gamma}, \tag{5.16}$$

$$Z(x) = \frac{V(x)}{I(x)} = Z_0 \frac{1+\Gamma \exp(-2j\beta x)}{1-\Gamma \exp(-2j\beta x)}. \tag{5.17}$$

Using $\Gamma = \frac{Z_l - Z_0}{Z_l + Z_0}$, we get

$$Z(x) = Z_0 \frac{Z_L \cos \beta x + j Z_0 \sin \beta x}{Z_0 \cos \beta x + j Z_L \sin \beta x}. \tag{5.18}$$

Until now, we assumed perfect conductors and the dielectric separating them as perfect insulator. In practice, conductor has small resistance R per unit length and dielectric has small admittance G per unit length.

Thus, we get two equations of the transmission line change to

$$-\frac{\partial V}{\partial x} = L \frac{dI(x)}{dt} + RI, \tag{5.19}$$

$$-\frac{\partial I}{\partial x} = C \frac{dV(x)}{dt} + GV. \tag{5.20}$$

With steady state,

$$-\frac{\partial V}{\partial x} = (j\omega L + R) I(x), \tag{5.21}$$

$$-\frac{\partial I}{\partial x} = (j\omega C + G) V(x). \tag{5.22}$$

See Fig. 5.2b, where we take load at right and voltage generator at left. Let us take x going from right to left. Then the above equations read

$$\frac{\partial V}{\partial x} = (j\omega L + R) I(x), \tag{5.23}$$

$$\frac{\partial I}{\partial x} = (j\omega C + G) V(x). \tag{5.24}$$

$$\frac{\partial^2 V}{\partial x^2} = \gamma^2 V(x), \tag{5.25}$$

where $\gamma = \sqrt{(j\omega L + R)(j\omega C + G)} = \alpha + j\beta$ and let $Z_0 = \sqrt{\frac{j\omega L + R}{j\omega C + G}}$.

To fix some typical numbers, $R = 0.01\,\Omega/\text{m}$, $L = 0.01\,\mu\text{H/m}$, $C = 100\,\text{pF/m}$, and $G = 0.01\,\mho/\text{m}$ giving at 2 GHz, $Z_0 = 10 + j0.04\,\Omega$.

$$V(x) = V_+ \exp(\gamma x) + V_- \exp(-\gamma x), \tag{5.26}$$

$$I(x) = \frac{1}{Z_0}(V_+ \exp(\gamma x) - V_- \exp(-\gamma x)) \tag{5.27}$$

$$= (I_+ \exp(\gamma x) - I_- \exp(-\gamma x)). \tag{5.28}$$

Then for $\Gamma = \frac{V_-}{V_+}$,

$$Z(0) = Z_L = \frac{V(0)}{I(0)} = Z_0 \frac{1+\Gamma}{1-\Gamma}, \tag{5.29}$$

$$Z(x) = \frac{V(x)}{I(x)} = Z_0 \frac{1+\Gamma \exp(-2\gamma x)}{1-\Gamma \exp(-2\gamma x)}. \tag{5.30}$$

Using $\Gamma = \frac{Z_l - Z_0}{Z_l + Z_0}$, we get

$$Z(x) = Z_0 \frac{Z_L \cosh \gamma x + Z_0 \sinh \gamma x}{Z_0 \cos \gamma x + Z_L \sinh \gamma x}. \tag{5.31}$$

For low loss transmission line where $\gamma = j\beta$, we get $Z(x)$ as in Eq. (5.18). There is no loss in the transmission line; then the power transferred to the load is as follows:

5.3 Impedance Matching and Smith Charts

Consider a voltage source V with source impedance Z_0. It is attached to transmission line with characteristic impedance Z_0 and a load at the end of transmission line Z_L as shown in Fig. 5.3. We want to know what should be the value of load impedance Z_L so that there is maximum power transferred to load.

Then voltage V_e at the end of the transmission line is

$$V_e = V_+ + V_- = V_+(1+\Gamma). \tag{5.32}$$

Then voltage V_b and current i_b at the beginning of the transmission line are

$$V_b = V_+ \exp(j\beta l) + V_- \exp(-j\beta l), \tag{5.33}$$

and

$$i_b = \frac{1}{Z_0}(V_+ \exp(j\beta l) - V_- \exp(-j\beta l)). \tag{5.34}$$

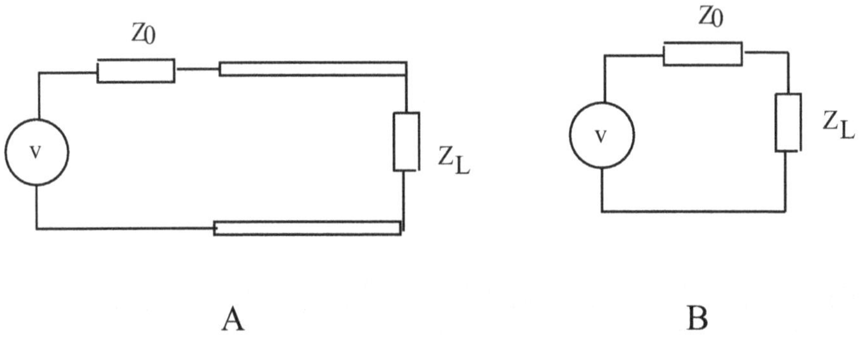

Fig. 5.3 (**a**) Voltage source V with impedance Z_0, a transmission line, and a load Z_L. (**b**) Load as seen from the generator end of the transmission line

$$V_b = V_0 - i_b Z_0, \tag{5.35}$$

giving

$$V_+ = \frac{V_0}{2} \exp(-j\beta l). \tag{5.36}$$

The power transferred is $P = Re(V_e I_e^*)$, which gives

$$P = \frac{|V_+|^2}{Z_0}(1 - |\Gamma_L|^2), \tag{5.37}$$

where

$$\Gamma = \frac{1 - \frac{Z_0}{Z_L}}{1 + \frac{Z_0}{Z_L}}. \tag{5.38}$$

For maximum transfer, $\Gamma = 0$ or $Z_L = Z_0$.

What if $Z_L \neq Z_0$, what should we do? We can move back from the load toward the voltage source. What we show is after some distance x, the resistance $Z(x)/Z_0$ becomes $1 + jv$. We can arrange to cancel the reactive part v by using an additional piece of transmission line, leaving the resulting load $Z(x) = Z_0$ which allows for maximum power transfer.

We can graphically find this distance x using a Smith chart. Given

$$r + jx = \frac{Z(x)}{Z_0} = \frac{1 + \Gamma}{1 - \Gamma}. \tag{5.39}$$

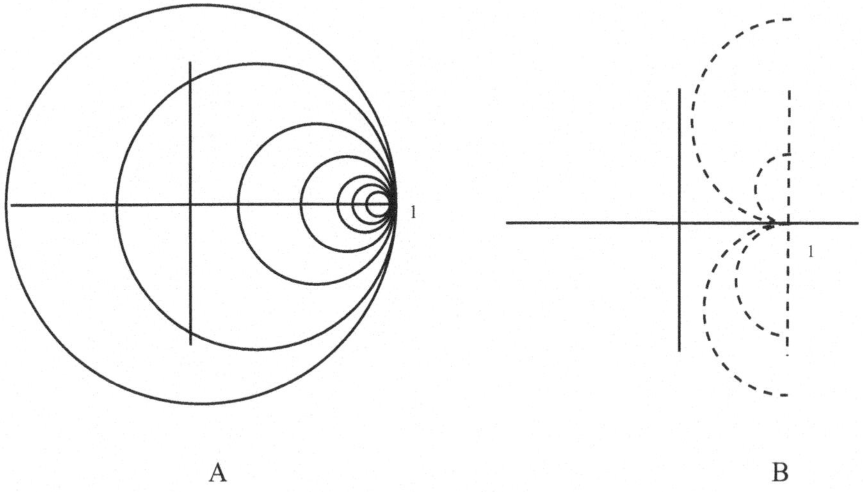

Fig. 5.4 Constant resistance (**a**) and reactance circle (**b**)

Then for $\Gamma = u + jv$, we can plot the locus of Γ for fixed r and fixed x. These are called constant resistance and reactance circles. These are given by

$$\left(u - \frac{r}{r+1}\right)^2 + v^2 = (r+1)^{-2} \tag{5.40}$$

a circle centered at $(\frac{r}{r+1}, 0)$ with radius $\frac{1}{r+1}$. This is called constant resistance circle. These are depicted in Fig. 5.4a, with smaller circle corresponding to larger r.

Similarly, constant reactance circle is

$$(u-1)^2 + \left(v - \frac{1}{x}\right)^2 = \frac{1}{x^2}, \tag{5.41}$$

which is centered at $(1, \frac{1}{x})$ with radius $\frac{1}{|x|}$. These are depicted in Fig. 5.4b, with smaller circle corresponding to larger x.

We start with a given load, corresponding to given (r, x). The intersection of constant r and x circles uniquely determines Γ. As we move back on transmission line, $\Gamma(x) \rightarrow \Gamma \exp(-j2\beta x)$, that is, we rotate Γ counterclockwise, keeping its magnitude constant, till it hits the constant resistance $r = 1$ circle. We stop and note the reactance value there, and this is the reactance we need to annul. All this is depicted in Fig. 5.5, where we located the normalized load $r + jx$ as the intersection of constant resistance circle r (shown in solid) and constant reactance circle x (shown in dashed) as point A. This point is rotated on circle centered at origin (shown in dotted) till we hit constant unit resistance circle (shown in bold) at

Fig. 5.5 Calculate using a Smith chart the distance traveled up a transmission line to reach unity normalized resistance

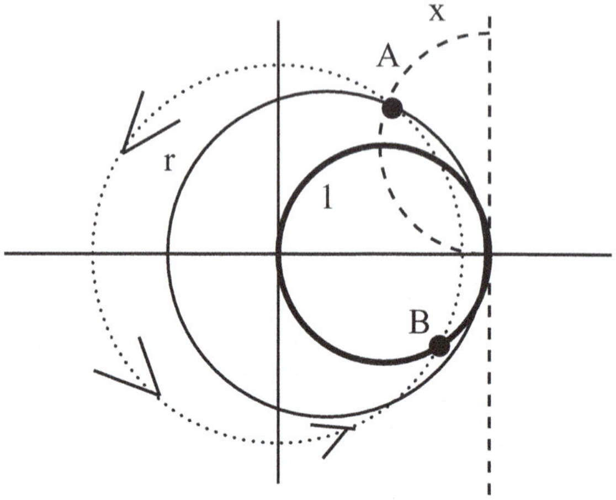

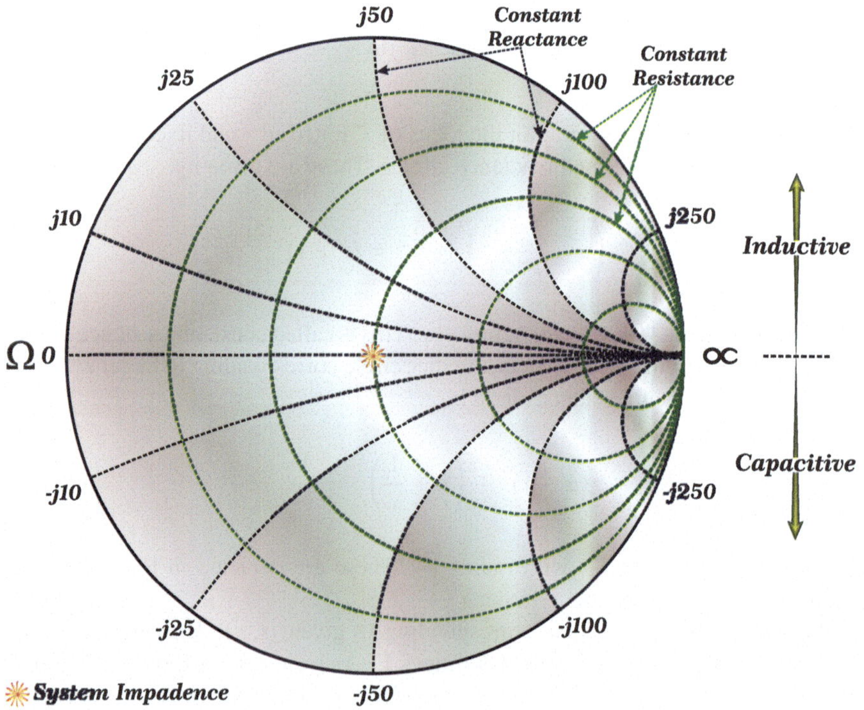

Fig. 5.6 Smith chart

point B. We can read the reactance of point B. The angle θ of rotation of A to B is given as $\Theta = 2\beta x$.

Figure 5.6 shows a Smith chart.

How do we cancel the reactance? One way is to add a segment transmission line in parallel (shunt) as shown in Fig. 5.7a.

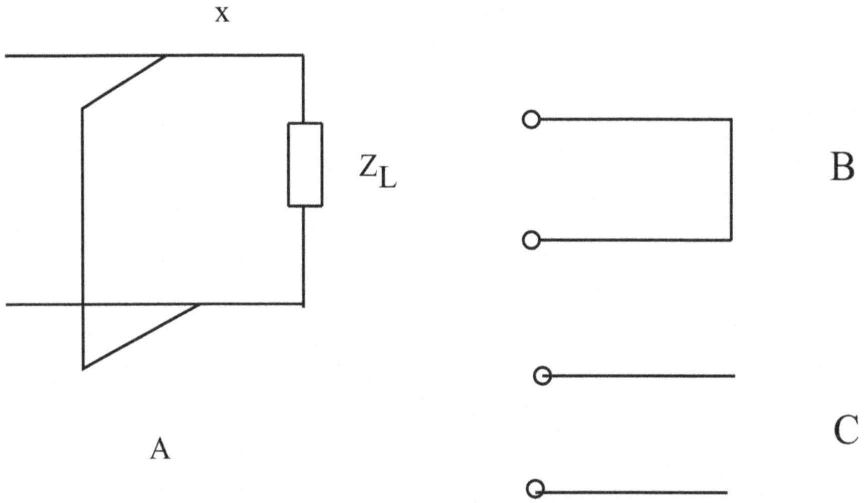

Fig. 5.7 (**a**) Impedance matching done by adding admittance in parallel to transmission line also called stubs. (**b**) Short circuited transmission line. (**c**) Open-circuited transmission line

The admittance of the shunt adds to admittance of the load at x. With $Y(x) = Z^{-1}(x)$ and $Y_0 = Z_0^{-1}$, we find for $\Gamma' = -\Gamma$; we have

$$g + jy = \frac{Y(x)}{Y_0} = \frac{1 + \Gamma'}{1 - \Gamma'}. \tag{5.42}$$

Now everything is same in the procedure as before. We start with a given load corresponding to admittance $g + jy$. The intersection of constant g and y circles uniquely determine Γ'. As we move back on transmission line $\Gamma'(x) \rightarrow \Gamma' \exp(-j2\beta x)$, that is, we rotate Γ' counterclockwise, keeping its magnitude constant, till it hits the constant conductance $g = 1$ circle. We stop and note the susceptance value there, and this is the susceptance we need to annul. All this is same as depicted in Fig. 5.5. Except now our starting point A is normalized admittance which is different from normalized impedance. The susceptance at point B is removed by adding a transmission line in shunt.

What is the impedance and admittance of a transmission line of length l. If it is closed at the other end as shown in Fig. 5.7b, then it means $Z_L = 0$ and from Eq. (5.18), $Z(l) = j \tan \beta l$ or $Y(l) = -j \cot \beta l$. If it is open at the other end as shown in Fig. 5.7c, then it means $Z_L = \infty$ and from Eq. (5.18), $Z(l) = -j \cot \beta l$ or $Y(l) = j \tan \beta l$. Therefore, to cancel a positive susceptance, we should use a closed-ended transmission line of $0 \leq l \leq \frac{\lambda}{4}$, and to cancel a negative susceptance, we should use an open-ended transmission line of $0 \leq l \leq \frac{\lambda}{4}$.

Observe $\Gamma(x)$ repeats every $2\beta x = 2\pi$, i.e., every $x = \frac{\lambda}{2}$. Therefore, load resistance is the same when we step back, $x = \frac{\lambda}{2}$. In fact, the maximum voltage on transmission

$$V_{\max} = V_{+}(1 + |\Gamma|), \tag{5.43}$$

and

$$V_{\min} = V_{+}(1 - |\Gamma|). \tag{5.44}$$

and the ratio $\frac{V_{\max}}{V_{\min}} = \frac{1+|\Gamma|}{1-|\Gamma|}$ is called voltage standing wave ratio (**VSWR**). When **VSWR** $= 1$, it implies $\Gamma = 0$, and there is no reflected wave. To check if we have matched the load perfectly, we measure **VSWR** to find if **VSWR** $= 1$.

5.4 Problems

1. For a coaxial cable of inner and outer diameter 1 and 2 cm, find inductance and capacitance per unit length.
2. Find the velocity of wave on the coax.
3. Find the wavelength if 10 GHz wave travels on this transmission line.
4. If a line is 1 m long, find the input impedance if the load impedance is open circuit and closed circuit.
5. Find the characteristic impedance for the line (you may neglect resistance per unit length).

Chapter 6
RF and Microwave Engineering

6.1 Microwave Oscillators

We first discuss how to produce microwaves (3–300 GHz) frequencies. These are produced in cavities, called microwave oscillators. The basic principle is to excite the modes of a cavity resonantly by motion of an electron. It is discussed in the following:

6.1.1 *Coaxial Cylindrical Cavity*

Consider a cylindrical cavity as shown in Fig. 6.1a. Figure 6.1a shows cross section. The cavity has an outer conductor of radius r_2 and an inner conductor of radius r_1 with $r_2 - r_1 = a$. The cavity supports both TE and TM modes. We discuss both.

6.1.2 *TE Modes*

In cylindrical coordinates (r, θ, z), we have

$$\frac{\partial}{\partial x} = \cos\theta \frac{\partial}{\partial r} - \frac{\sin\theta}{r}\frac{\partial}{\partial \theta}, \tag{6.1}$$

$$\frac{\partial}{\partial y} = \sin\theta \frac{\partial}{\partial r} + \frac{\cos\theta}{r}\frac{\partial}{\partial \theta}. \tag{6.2}$$

For TE modes from Eqs. (3.13 and 3.14), we can write

N. Khaneja, *Electromagnetic Waves and Optics*,
https://doi.org/10.1007/978-3-031-76144-7_6

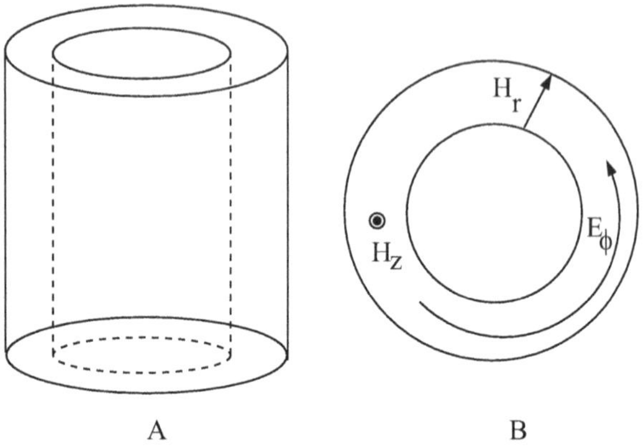

Fig. 6.1 (**a**) Cylindrical cavity with inner and outer conductors. (**b**) Its cross section containing TE mode

$$j\omega\left(\epsilon - \frac{\beta^2}{\mu\omega^2}\right) E_r = \frac{1}{r}\frac{\partial H_z}{\partial \theta}, \tag{6.3}$$

$$j\omega\left(\epsilon - \frac{\beta^2}{\mu\omega^2}\right) E_\theta = -\frac{\partial H_z}{\partial r}. \tag{6.4}$$

From Eqs. (3.15 and 3.16), we get

$$\frac{\mu}{\beta} j\omega^2\left(\epsilon - \frac{\beta^2}{\omega^2\mu}\right) H_r = \frac{\partial H_z}{\partial r}, \tag{6.5}$$

$$\frac{\mu}{\beta} j\omega^2\left(\epsilon - \frac{\beta^2}{\omega^2\mu}\right) H_\theta = \frac{1}{r}\frac{\partial H_z}{\partial \theta}. \tag{6.6}$$

We choose

$$H_z = f(r)\cos n\theta \tag{6.7}$$

Then we get

$$-\underbrace{\left(\frac{\omega^2}{c^2} - \beta^2\right)}_{l^2} H_z = \left(\frac{\partial^2}{\partial r^2} + \frac{1}{r}\frac{\partial}{\partial r} + \frac{1}{r^2}\frac{\partial^2}{\partial \theta^2}\right) H_z.$$

$$\left(-l^2 + \frac{n^2}{r^2}\right) f(r) = \left(\frac{\partial^2}{\partial r^2} + \frac{1}{r}\frac{\partial}{\partial r}\right) f(r).$$

The solution in the annulus is $J_n(lr + o)$. If $Z_1, Z_2, \ldots, Z_n$ are zeros of the $J_n'(x)$ (other than trivial zeros at 0), then mode lm is defined as

$$lr_1 + o = Z_l, \tag{6.8}$$

$$lr_2 + o = Z_m. \tag{6.9}$$

This ensures that E_θ and H_r vanish at the boundary. We solve for $l = \frac{Z_m - Z_l}{a}$.

This defines $\beta^2 = \sqrt{(\frac{\omega^2}{c^2} - (\frac{Z_m - Z_l}{a})^2}$. We have a wave along z with dependence $\exp(-j\beta z)$. Superimposing on it a reflected wave $\exp(j\beta z)$, we get transverse electric field has z dependence $\sin(\beta z)$ and transverse magnetic field $\cos(\beta z)$. By choosing $\beta = \frac{p\pi}{b}$, we get a standing wave mode indexed by p and Z_m, Z_l, with

$$\omega = c\sqrt{\left(\frac{p\pi}{b}\right)^2 + \left(\frac{Z_m - Z_l}{a}\right)^2}. \tag{6.10}$$

6.2 TM Modes

For TM modes from Eqs. (3.28 and 3.29), we can write

$$j\omega\left(\mu - \frac{\beta^2}{\epsilon\omega^2}\right)H_r = \frac{1}{r}\frac{\partial E_z}{\partial\theta}, \tag{6.11}$$

$$j\omega\left(\mu - \frac{\beta^2}{\epsilon\omega^2}\right)H_\theta - \frac{\partial E_z}{\partial r}. \tag{6.12}$$

From Eqs. (3.26 and 3.27), we get

$$\frac{\epsilon}{\beta}j\omega^2\left(\mu - \frac{\beta^2}{\omega^2\epsilon}\right)E_r = \frac{\partial E_z}{\partial r}, \tag{6.13}$$

$$\frac{\epsilon}{\beta}j\omega^2\left(\mu - \frac{\beta^2}{\omega^2\epsilon}\right)E_\theta = \frac{1}{r}\frac{\partial E_z}{\partial\theta}. \tag{6.14}$$

We choose

$$E_z = f(r)\cos n\theta. \tag{6.15}$$

$$j\omega\epsilon E_z = \frac{\partial E_y}{\partial x} - \frac{\partial E_x}{\partial y} \propto \frac{\partial^2 E_z}{\partial x^2} + \frac{\partial^2 E_z}{\partial y^2},$$

which gives

From

$$\underbrace{\left(\frac{\omega^2}{c^2}-\beta^2\right)}_{l^2}E_z=\left(\frac{\partial^2}{\partial r^2}+\frac{1}{r}\frac{\partial}{\partial r}++\frac{1}{r^2}\frac{\partial^2}{\partial\theta^2}\right)E_z.$$

$$\left(-l^2+\frac{n^2}{r^2}\right)f(r)=\left(\frac{\partial^2}{\partial r^2}+\frac{1}{r}\frac{\partial}{\partial r}\right)f(r).$$

The solution in the annulus is $J_n(lr+o)$. If $Y_1, Y_2, \ldots, Y_n$ are zeros of the $J_n(x)$ (other than trivial zeros at 0), then mode lm is defined as

$$lr_1+o=Y_l, \tag{6.16}$$

$$lr_2+o=Y_m \tag{6.17}$$

This ensures that E_θ and H_r vanish at the boundary. We solve for $l=\frac{Y_m-Y_l}{a}$.

This defines $\beta^2=\sqrt{(\frac{\omega^2}{c^2}-(\frac{Y_m-Y_l}{a})^2}$. We have a wave along z with dependence $\exp(-j\beta z)$. Superimposing on it a reflected wave $\exp(j\beta z)$, we get transverse electric field has z dependence $\sin(\beta z)$ and transverse magnetic field $\cos(\beta z)$. By choosing $\beta=\frac{p\pi}{b}$, we get a standing wave mode indexed by p and Y_m, Y_l, with

$$\omega=c\sqrt{\left(\frac{p\pi}{b}\right)^2+\left(\frac{Y_m-Y_l}{a}\right)^2}. \tag{6.18}$$

6.2.1 Magnetron Oscillator

In Fig. 6.2a, we have a coaxial cylindrical cavity as shown. Consider TE mode ϕ that has angular dependence $\cos\theta$ and frequency ω_0. If we introduce an electron in the cavity and a magnetic field B_0 along the z direction, then this electron will press around the magnetic field at the angular velocity $\omega=\frac{eB}{m}$. If we choose B_0 such that $\omega=\omega_0$, then the electron orbits in the cavity at angular frequency ω_0. This comprises a circulating current density that is nonzero at angular position $\theta=\omega t$, and hence its overlap with the TE mode (E_ϕ part of it) is of the form $\cos\omega_0 t$. Thus, we have a mode at frequency ω_0, and we excite it as $\cos\omega_0 t=\frac{\exp(j\omega_0 t)+\exp(-j\omega_0 t)}{2}$. Thus, we excite a linear system at resonance, and we will build up this mode ϕ. Rapidly, the cavity will be filled up with this mode, and if ω_0 is in the GHz range, we produce microwaves.

We discussed the TE mode, but we also have the TM modes. Let ϕ_1 be a TM mode that has angular dependence $E_\phi \propto \sin\theta$ and some frequency ω_0. If we

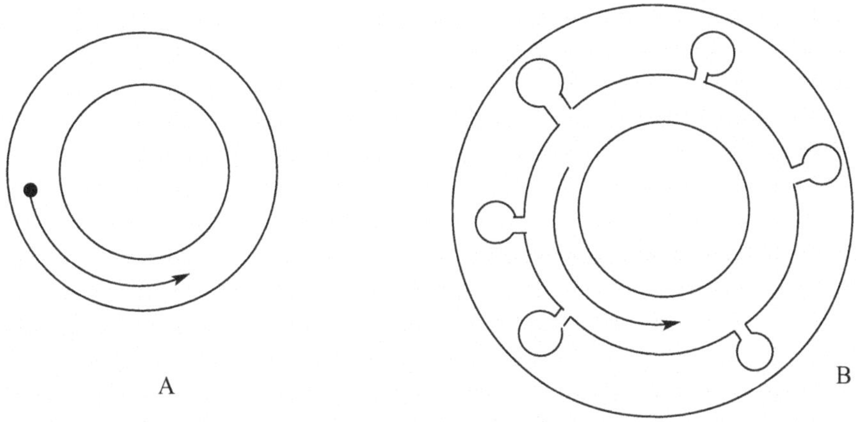

Fig. 6.2 (**a**) Cross section of a cylindrical cavity with electron circulating the space between inner and outer conductor. (**b**) Inner cylindrical cavity aperture coupled to outer cavities with the same resonant frequency

introduce an electron in the cavity and a magnetic field B_0 along z direction, then this electron will press around the magnetic field at the angular velocity $\omega = \frac{eB}{m}$. If we choose B_0 such that $\omega = \omega_0$, then the electron orbits in the cavity at angular frequency ω_0. This comprises a circulating current density that is nonzero at angular position $\theta = \omega t$, and hence its overlap with the TM mode (E_ϕ part of it) is of the form $\sin \omega_0 t$. Thus, we have a mode at frequency ω_0, and we excite it as $\sin \omega_0 t = \frac{\exp(j\omega_0 t) - \exp(-j\omega_0 t)}{2j}$. Thus, we excite a linear system at resonance, and we will build up this mode ϕ. Rapidly, the cavity will be filled up with this mode, and if ω_0 is in the GHZ range, we produce microwaves.

Thus, we are saying that if the precession frequency of electron matches the frequency of either the TE or TM mode with $\cos\theta$ dependence, we will excite these modes. We also have modes with angular dependence $\cos n\theta$ and frequency ω_0. Then it is excited with electron with precession frequency $\frac{\omega_0}{n}$. The electron is injected into cavity from inner conductor that acts as cathode. The electrons are injected radially outward and under the Lorentz force of the magnetic field acquire a circular orbit as shown in Fig. 6.2a. The strength of magnetic field determines the angular velocity ω as $\omega = \frac{eB}{m}$.

Figure 6.2b shows the image of a microwave oscillator called magnetron. The inner conductor is a cathode that injects electrons. The central conductor is an anode that attracts these electrons radially and makes them acquire a velocity which under magnetic field gets turned into a circular orbit. Between the center conductor and outer conductor are circular chambers/cavities which resonate at frequency ω_0 and can store microwaves. Without these cavities, the volume of the central cavity is small, and as we fill the cavity with microwaves, it offers increased resistance to motion of electron. Observe that the circulating electron creates mode $-\phi$ and the resulting E_ϕ opposes electron motion. By adding more room, in terms of outer cavities, we store the same energy at a lower E_ϕ value and reduced resistance. Thus, we have an arrangement where we have central cavity as shown in Fig. 6.2a aperture coupled to outer cavities all of which resonate at the same frequency. We excite the central cavity by cyclotron motion of electron, and the outer cavities provide more storage room.

6.2.2 Klystron and Gyrotron

In magnetron, we have a cylindrical mode with cos $theta$ dependence and frequency ω_0. We traverse the mode with an electron rotating circularly at angular frequency ω_0, which excites the mode resonantly as $\cos \omega_0 t$. Now consider the cylindrical cavity again (without central conductor). We have TM modes in the cavity with E_z having $\cos kz$ dependence. If we make an electron travel along the z direction with velocity v as shown in Fig. 6.3, then we are exciting this mode as $\sin kvt$ and with $\omega = kv$, we can write the excitation as $\sin \omega t$. Now if ω matches ω_0 the resonant frequency of the mode, then we have excited the mode with an rectilinear motion. This is the principal of a klystron, a device used to produce microwaves especially for the radar application.

In magnetron, we made the electron move in circles which talked to $\cos \theta$ dependence of the mode. In klystron, we made the electron move in the z direction and it talked to $\cos kz$ dependence of the mode. Now we can do both. We can make the electron gyrate as shown in Fig. 6.3b, where we travel vertically along z and also in loops. The loopy motion comes from an applied magnetic field. If ω_1 is the angular (loopy) velocity of the electron and electron travels along the z direction with velocity v, such that $kv = \omega_2$. Then we have excitation of the mode as $\cos \omega_1 t \cos \omega_2 t$, i.e., we excite the mode at frequency $\omega_1 \pm \omega_2$. If the frequency of mode is say $\omega_0 = \omega_1 + \omega_2$, we obtain resonant excitation of the mode. Furthermore, as we add two frequencies $\omega_1 + \omega_2$, we can get a high resulting frequency ω_0. This mechanism of producing microwaves by gyrating motion is realized in a device called gyrotron.

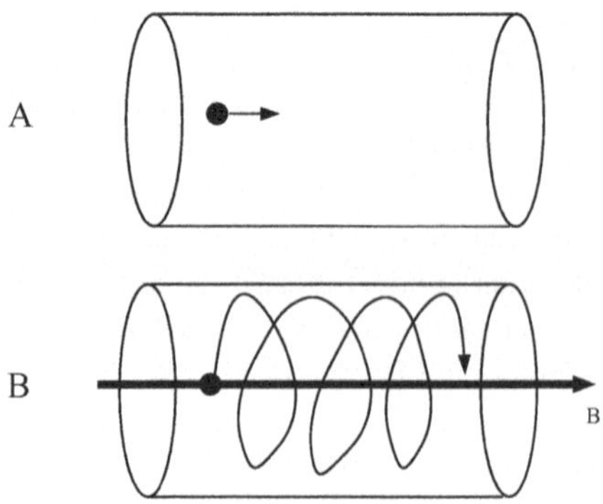

Fig. 6.3 Cylindrical cavity being resonantly excited by linear and gyrating motion of electron

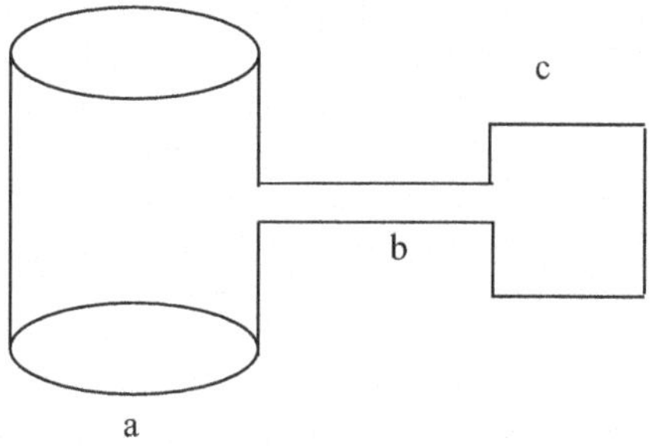

Fig. 6.4 Microwave network with source and load cavity connected by a waveguide

6.2.3 *Microwave Systems and Networks*

We have discussed microwave cavities and waveguides. Now we discuss microwave networks made out of these systems. See Fig. 6.4, which shows a cylindrical cavity (source cavity) acting as a microwave source (a) coupled to a waveguide (b), which is terminated in a microwave cavity (load cavity) (c), which could be a cavity as in microwave oven in which we heat our food. The modes of the source are appropriately excited, and we build these modes. The EM radiation is then channeled through a waveguide to resonator. How to analyze such networks?

Let us say source cavity has a mode ϕ_0 at frequency ω_0. Let us normalize the mode such that $\langle \phi_0, \phi_0 \rangle = 1$, $(\langle \phi_0, \phi_0 \rangle = \int \epsilon |E|^2 + \mu |H|^2)$. When we excite the source cavity mode resonantly, its amplitude x builds up linearly as time as $\dot{x} = u$. Now suppose we make a small aperture of area A_e in the wall of the source cavity to let microwaves out. We can analyze the aperture by assuming that there is no aperture, in which case the mode will cause certain current on the walls of the waveguide, and now we subtract this current by assuming a fictitious current i_0, opposite to mode current on the wall. This is shown in Fig. 6.5a. This fictitious current i_0 radiates inside and outside the cavity. The radiation inside produces H in proximity to current that opposes H due to mode current, and hence the fictitious current radiates inward to de-excite the mode. Since fictitious mode is proportional to mode amplitude, we can write

$$\dot{x} = -\alpha_0^2 x + u. \tag{6.19}$$

How much is α^2? Let E be the electric field value of ϕ_0 at the aperture. If the wall was real, this E will exponentially decay in the wall and penetrate length l inside the wall. The inner product of current $j = \sigma E$ with this ϕ_0 gives

$$\alpha_0^2 = \frac{cA}{V}, \tag{6.20}$$

where A is the aperture area, V is mode volume, and c is light velocity,

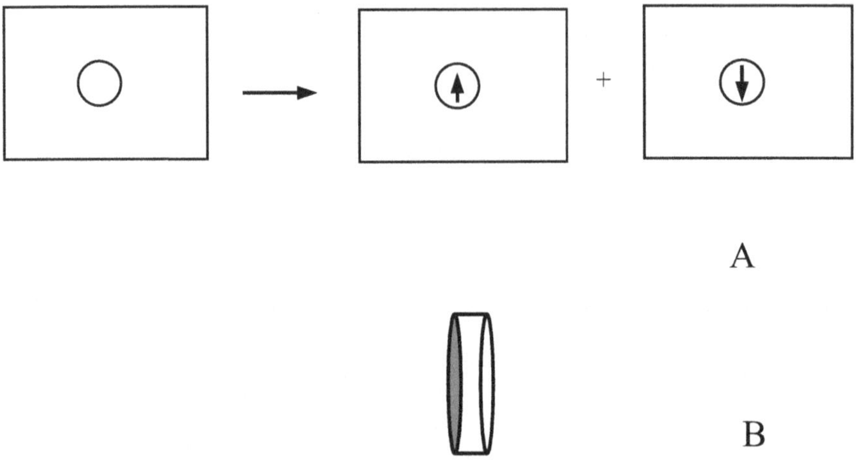

Fig. 6.5 (**a**) Aperture modeled as mode current with no aperture and a fictitious current that opposes this current. (**b**) Electric field in infinitesimal volume inside the wall where it exponentially decays

$$\dot{x} = -\alpha_0^2 x + u, \tag{6.21}$$

So we excite the mode, but it leaks out of aperture and decays at rate α_0^2. We can solve the above equation, and we get steady amplitude of mode in the cavity as $x = \frac{u}{\alpha_0^2}$.

Forget loss for a moment. What if we excited the cavity away from resonance frequency ω_0 at ω? Then we can describe the mode amplitude at frequency $s = j\omega$ as $x(s) = \frac{1}{s-a}u(s)$, where $a = j\omega_0$. The quantity $G(s) = \frac{1}{s-a}$ is called the cavity transfer function. At $s = j\omega_0$, it becomes infinite as on resonance, the cavity amplitude grows unbounded. This is shown in Fig. 6.6a. In the presence of loss now, the transfer function is modified as feedback as shown in Fig. 6.6b and the new transfer function $G_1(s)$ is

$$G_1(s) = \frac{G(s)}{1 + \alpha^2 G(s)} = \frac{1}{s - a + \alpha^2}. \tag{6.22}$$

Now we connect the source cavity to the waveguide which is coupled to load cavity. How do we analyze this interconnect? Let us simplify and say we have two resonant cavities connected by a small aperture as shown in Fig. 6.7a.

Let ϕ_A be the mode of cavity A and ϕ_B be the mode of cavity B both with frequency ω_0 before we create an aperture. Both modes will create currents on cavity wall. If modes are say identical, then ϕ_A and ϕ_B will create the same current where aperture is. If we consider $\phi = \phi_A - \phi_B$, the currents due to ϕ_A and ϕ_B will cancel and ϕ has no current at the boundary where aperture is, and we get a mode of the cavity with aperture.

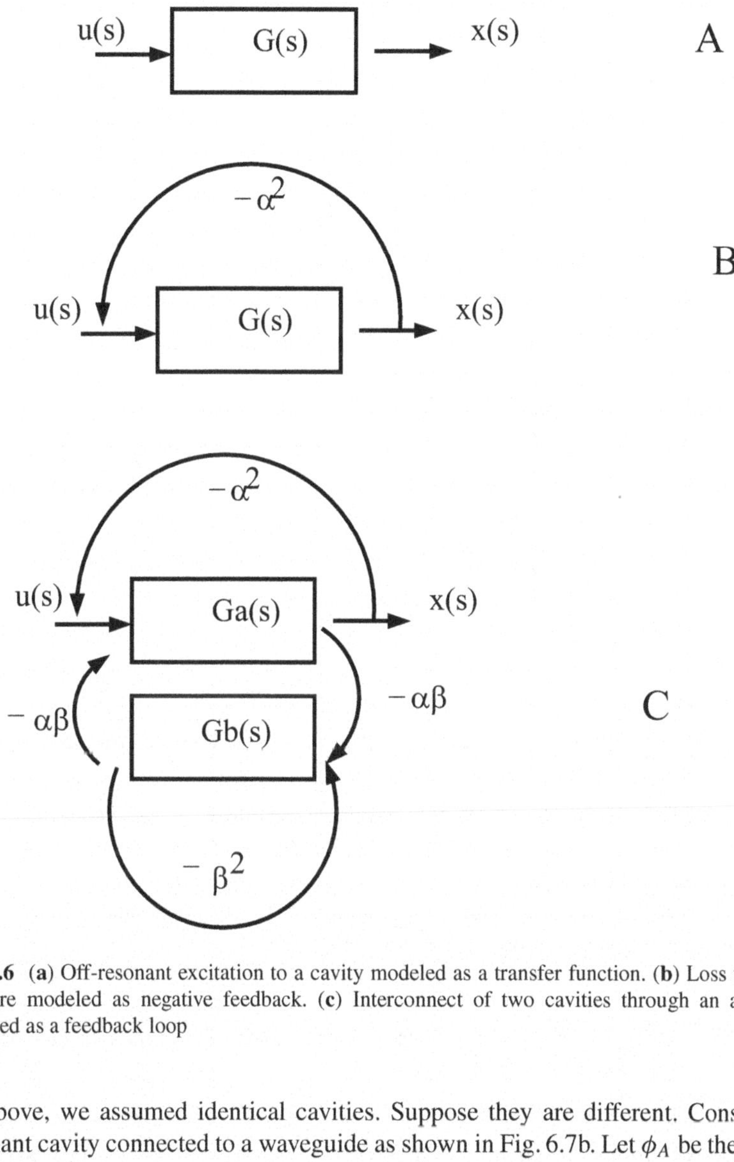

Fig. 6.6 (**a**) Off-resonant excitation to a cavity modeled as a transfer function. (**b**) Loss through aperture modeled as negative feedback. (**c**) Interconnect of two cavities through an aperture modeled as a feedback loop

Above, we assumed identical cavities. Suppose they are different. Consider a resonant cavity connected to a waveguide as shown in Fig. 6.7b. Let ϕ_A be the mode of cavity and ϕ_B be the mode of waveguide both with frequency ω_0 (say both modes normalized). The aperture current distribution due to ϕ_A and ϕ_B are different in general. Let us close cavity A by introducing fictitious current i_A and close cavity B by introducing fictitious current i_B. If x and y are the amplitudes of modes ϕ_A and ϕ_B, we can write

$$\frac{d}{dt}\begin{bmatrix} x \\ y \end{bmatrix} = \begin{bmatrix} -\alpha_0^2 & -\alpha\beta \\ -\alpha\beta & -\beta^2 y \end{bmatrix}\begin{bmatrix} x \\ y \end{bmatrix} + \begin{bmatrix} u \\ 0 \end{bmatrix}. \tag{6.23}$$

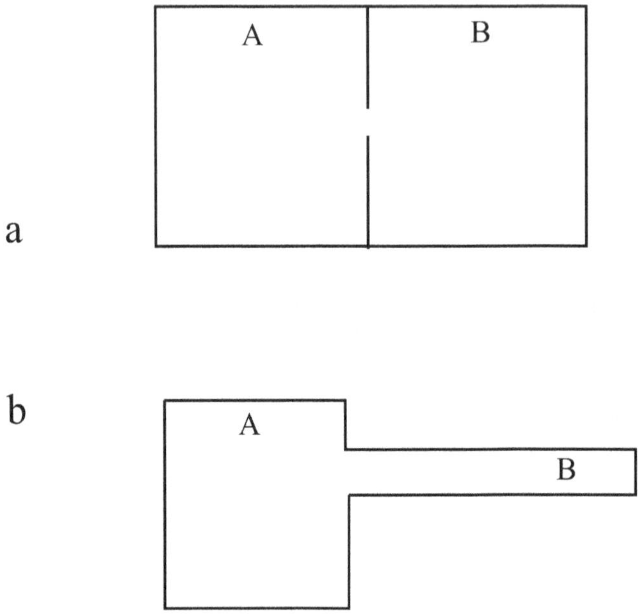

Fig. 6.7 (**a**) and (**b**) Aperture coupling of two identical and different cavities

where $\alpha = \sqrt{\frac{cA}{V_1}}$ and $\beta = \sqrt{\frac{cA}{V_2}}$ with V_1 and V_2 as cavity volumes for cavities 1 and 2.

Since the two cavities have different volumes, we find $\alpha^2 \neq \beta^2$. From Eq. (6.23), we find the smaller cavity has a larger loss. Furthermore, $\beta\phi_A - \alpha\phi_B$ is the joint mode that does not dissipate. When we excite, this mode builds up. Now suppose the two cavities do not share the same resonance. Let the resonances be a and b. Then the joint transfer function as shown in Fig. 6.6c is

$$G(s) \propto \frac{1}{1 - \frac{\alpha^2\beta^2}{(s-a+\alpha^2)(s-b+\beta^2)}} \propto \frac{1}{(s-a)(s-b) + \beta^2(s-a) + \alpha^2(s-b)} \tag{6.24}$$

Now see two limits, when $\alpha \gg \beta$, $G(s) \propto \frac{1}{(s-a)(s-b)+\alpha^2(s-b)}$ i.e., we have resonance at a. When $\alpha \ll \beta$, we have $G(s) \propto \frac{1}{(s-a)(s-b)+\beta^2(s-a)}$ i.e., we have resonance at b. For intermediate values, it is a weighted sum. If we assume that the cavity volume is much larger than the waveguide volume, then the final resonance is more weighted toward cavity resonance. If both waveguide and cavity have resonance at a then as expected the joint system gas has resonance at a irrespective of $\alpha and \beta$ values.

Now return to Fig. 6.4. Suppose source and cavity have resonance at a and suppose waveguide is off, may be its length is not correct. The length is decided between where source and destination are and not much can be done then. Then the joint resonance of waveguide and load cavity will be different, and we do not do a good job of exciting this system by source. How can we change the waveguide resonance without changing its length so that everything is on resonance, i.e., impedance matched? This is called impedance matching.

Other scenario is that the load cavity resonance goes off a bit because we put something in it. Then we can change the waveguide resonance so that jointly they come back to the source resonance. In either case, we have to study how to vary the resonance of the waveguide. There are many ways to do it, and we study them under impedance matching.

6.3 Impedance Matching Waveguides

6.3.1 Taper

Recall that with a waveguide with TE_{mn} mode, we have

$$\beta = \sqrt{\frac{\omega^2}{c^2} - \left(\frac{m\pi}{a}\right)^2 - \left(\frac{n\pi}{b}\right)^2}. \tag{6.25}$$

Let us focus on TE_{11} mode with

$$\beta = \sqrt{\frac{\omega^2}{c^2} - \frac{\pi^2}{a^2} - \frac{\pi^2}{b^2}}. \tag{6.26}$$

If l is the length of the waveguide and if $\beta l = p\pi$, then the waveguide is resonant with ω. If $(p-1)\pi < \beta l < p\pi$, then we can increase the length and bring waveguide into resonance with ω. l is fixed so we can change the dimension of waveguide and make say a and b as functions of z as shown in Fig. 6.8, where in Fig. 6.8a we change b as a function of z and in Fig. 6.8b we change a as a function of z. Then

$$\beta(z) = \sqrt{\frac{\omega^2}{c^2} - \frac{\pi^2}{a^2(z)} - \frac{\pi^2}{b^2(z)}}. \tag{6.27}$$

As we increase say $a(z)$, β increases and now $\int_0^l \beta(z)dz = p\pi$, and we bring waveguide into resonance. Of course, this requires changing the structure of waveguide permanently. We prefer instead to not do permanent change to the

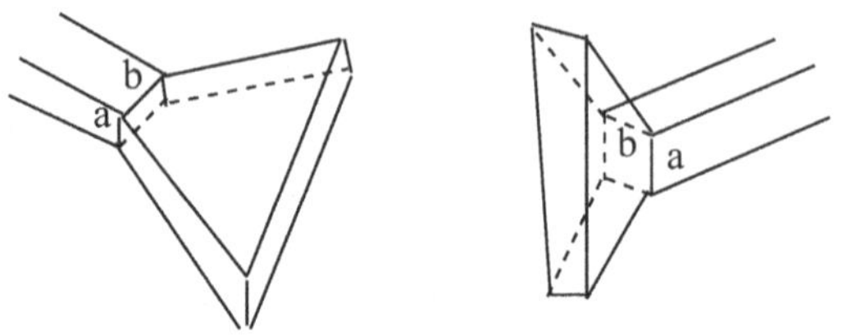

Fig. 6.8 (**a**) Changing b dimension of waveguide. (**b**) Changing a dimension of waveguide

waveguide structure and instead introduce alterations that change the impedance as described below. For this, we first develop a transmission line model of waveguide.

6.3.2 Transmission Line Models

To fix ideas, consider a waveguide with TE_{11} mode. The cross section of the waveguide shows electric field E_y in the y direction and E_x in the x direction as in Fig. 6.9a, b, respectively. The electric field deposits charges on the boundary and the cross section acts as capacitor. Let us see the field E_y; it has the form $E_0 \sin(\frac{\pi x}{a}) \cos(\frac{\pi y}{b})$ and it is simply $E_0 \sin(\frac{\pi x}{a})$ at the top and bottom boundaries, which gives a charge density $\frac{\sigma}{2\epsilon} = E_y$. We can integrate the charge density to get total charge Q, and then using the fact that energy stored in electric field $\int \epsilon_0 E_y^2 = \frac{Q^2}{C}$, we get the capacitance as $\epsilon_0 \frac{a \Delta x}{\pi b}$, where Δx is the small length in the z direction, or capacitance per unit length is $C_1 = \epsilon_0 \frac{a}{\pi b}$. Similarly, E_x gives $C_2 = \epsilon_0 \frac{b}{\pi a}$. We remark if we have TE_{mn} mode instead of TE_{11} then the capacitance $C_1 = \epsilon_0 \frac{a}{\pi m b}$ and $C_2 = \epsilon_0 \frac{b}{\pi n a}$.

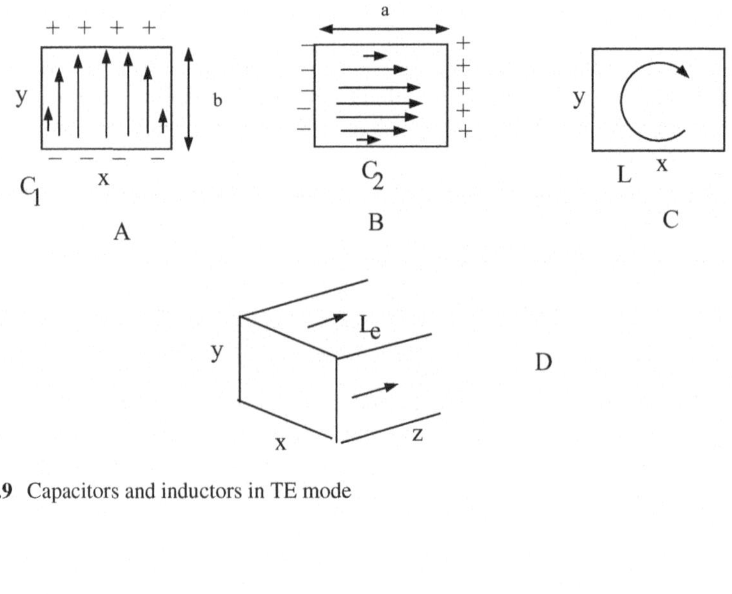

Fig. 6.9 Capacitors and inductors in TE mode

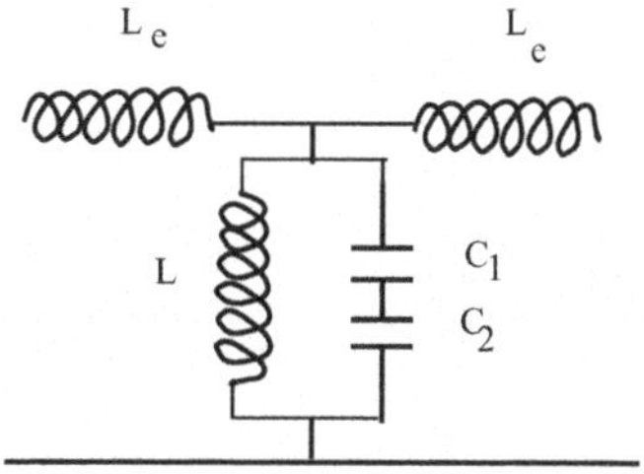

Fig. 6.10 Transmission line model for TE mode

The energy oscillates between electric field E_x, E_y and magnetic fields. The magnetic field H_z corresponds to circular currents as shown in Fig. 6.9c, which gives inductance L; the value of the inductance is just the area of the loop $\mu_0 ab$. For thickness Δx, we get $L \rightarrow \frac{L}{\Delta x}$. H_z is not the only magnetic field, but we also have magnetic field H_x, H_Y, which correspond to longitudinal currents as shown in Fig. 6.9d. Let us see the field H_y; it has the form $H_0 \sin(\frac{\pi y}{b}) \cos(\frac{\pi x}{z})$, and it is simply $H_0 \sin(\frac{\pi y}{b})$ at the side wall, which gives a current density $\frac{j}{2} = H_y$. We can integrate the current density to get total current I, and then using the fact that energy stored in the magnetic field $\int \mu_0 H_y^2 = LI^2$, we get the inductance as $\mu_0 \frac{a\Delta x}{\pi b}$, where Δx is the small length in the z direction, or inductance per unit length is $L_1 = \mu_0 \frac{\pi a}{b}$. Similarly, H_x gives $L_2 = \mu_0 \frac{\pi b}{a}$. We remark if we have TE_{mn} mode instead of TE_{11}, then the inductance $L_1 = \mu_0 \frac{\pi an}{b}$ and $L_2 = \mu_0 \frac{\pi bm}{a}$. Let

$$L_e = L_1 + L_2. \tag{6.28}$$

Then we have a transmission line model as shown in Fig. 6.10, with line length signifying the longitudinal z direction. Then we can write transmission line equations

$$-j\beta V(z) = j\omega L_e I(z), \tag{6.29}$$

$$-j\beta I(z) = \left(j\omega C_e + \frac{1}{j\omega L} \right) V(z), \tag{6.30}$$

where $C_e^{-1} = C_1^{-1} + C_2^{-1}$. On multiplying the above two equations, this gives

$$L_e C_e = c^{-2}. \tag{6.31}$$

$$\omega^2 = c^2\beta^2 + \frac{1}{LC_e} = c^2\beta^2 + \frac{1}{L}\left(\frac{1}{C_1} + \frac{1}{C_2}\right), \tag{6.32}$$

$$\frac{\omega^2}{c^2} = \beta^2 + \left(\frac{\pi}{a}\right)^2 + \left(\frac{\pi}{b}\right)^2. \tag{6.33}$$

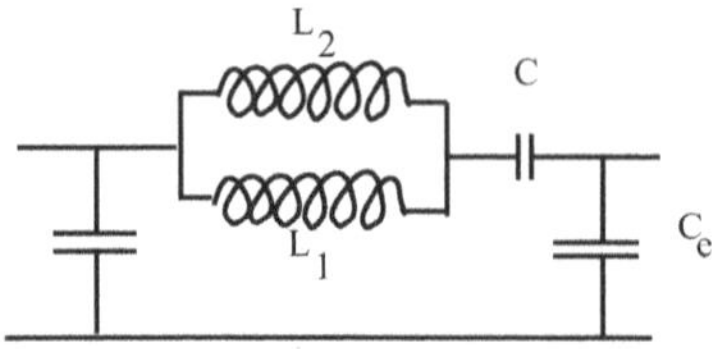

Fig. 6.11 Transmission line model for TM mode

or

$$\beta = \sqrt{\frac{\omega^2}{c^2} - \left(\frac{\pi}{a}\right)^2 - \left(\frac{\pi}{b}\right)^2}. \tag{6.34}$$

We now show we can change the capacitance of the cross section and get

$$\beta(z) = \frac{1}{c}\sqrt{\omega^2 - \frac{1}{L}\left(\frac{1}{C_1} + \frac{1}{C_2}\right)}. \tag{6.35}$$

There are various ways to change the capacitance of cross sections; we study them below.

Before we move forward, we develop a transmission line model for TM mode.

To fix ideas, consider a waveguide with TM_{11} mode. The cross section of the waveguide shows magnetic field H_y in the y and H_x in the x direction. These fields give longitudinal currents on side wall and top wall of the waveguide. The energy oscillates between magnetic field H_x, H_y and the electric fields. The electric field E_z corresponds to charge along longitudinal direction. These currents and charge along longitudinal direction are represented by two inductors L_1 and L_2 (for side and top walls, respectively) in parallel and a series capacitor C as shown in Fig. 6.11. L_1 and L_2 are the same as in TE mode. The capacitance $C = \frac{\epsilon_0 ab}{\Delta}$. The transverse electric fields deposit charge on top and side walls of waveguide cross section and are represented by capacitor

$$C_e = C_1 + C_2, \tag{6.36}$$

where C_1 and C_2 are the same as in the TE mode.

Then we have a transmission line model as shown in Fig. 6.11, with line length signifying the longitudinal z direction. Then we can write transmission line equations

$$-j\beta V(z) = \left(j\omega L_e + \frac{1}{j\omega C}\right) I(z), \tag{6.37}$$

$$-j\beta I(z) = j\omega C_e V(z), \tag{6.38}$$

where $L_e^{-1} = L_1^{-1} + L_2^{-1}$. This gives with $C_e L_e = c^{-2}$.

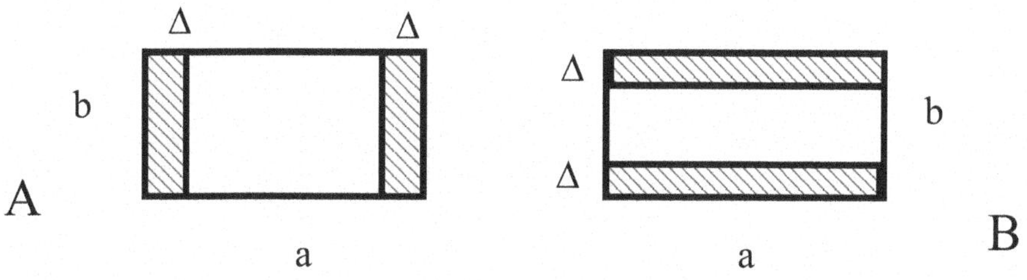

Fig. 6.12 Changing cross section of a waveguide using metal strip called iris

$$\omega^2 = c^2\beta^2 + \frac{1}{CL_e} = c^2\beta^2 + \frac{1}{C}\left(\frac{1}{L_1} + \frac{1}{L_2}\right), \tag{6.39}$$

$$\frac{\omega^2}{c^2} = \beta^2 + \left(\frac{\pi}{a}\right)^2 + \left(\frac{\pi}{b}\right)^2, \tag{6.40}$$

or

$$\beta = \sqrt{\frac{\omega^2}{c^2} - \left(\frac{\pi}{a}\right)^2 - \left(\frac{\pi}{b}\right)^2}. \tag{6.41}$$

6.3.3 Iris

Now focus on TE modes.

We change the cross section as shown in Fig. 6.12a. The two capacitance C_1 and C_2 change from $\frac{\epsilon_0 a}{b}$ and $\frac{\epsilon_0 b}{a}$ to $\frac{\epsilon_0(a-\Delta)}{b}$ and $\frac{\epsilon_0 b}{(a-\Delta)}$, respectively, and we get

$$\beta = \sqrt{\frac{\omega^2}{c^2} - \left(\frac{\pi}{a}\right)^2 - \left(\frac{\pi}{b}\right)^2 + \frac{\Delta}{a}\left(\left(\frac{\pi}{a}\right)^2 - \left(\frac{\pi}{b}\right)^2\right)}. \tag{6.42}$$

Since $a > b$, we find that β decreases.

We change the cross section as shown in Fig. 6.12b. The two capacitance C_1 and C_2 change from $\frac{\epsilon_0 a}{b}$ and $\frac{\epsilon_0 b}{a}$ to $\frac{\epsilon_0 a}{b-\Delta}$ and $\frac{\epsilon_0(b-\Delta)}{a}$, respectively, and we get

$$\beta = \sqrt{\frac{\omega^2}{c^2} - \left(\frac{\pi}{a}\right)^2 - \left(\frac{\pi}{b}\right)^2 - \frac{\Delta}{b}\left(\left(\frac{\pi}{a}\right)^2 - \left(\frac{\pi}{b}\right)^2\right)}. \tag{6.43}$$

Since $a > b$, we find that β increases.

Thus, using iris as above, we can increase or decrease β for a cross section and impedance match.

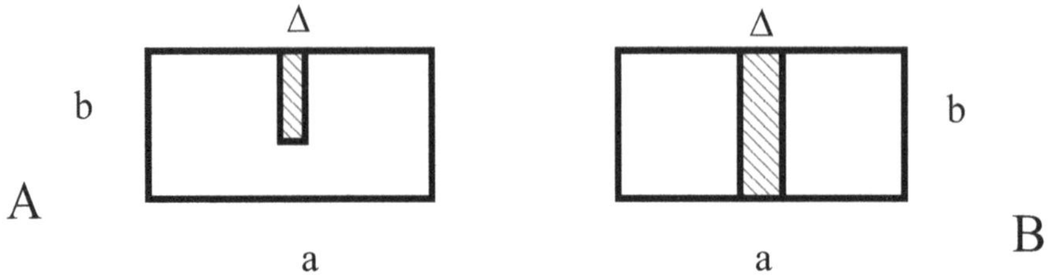

Fig. 6.13 Changing cross section of a waveguide using a tuning screw

6.3.4 Post and Screws

Focus on TE modes.

In addition to using a waveguide iris, post or screw can also be used to give a similar effect and thereby provide waveguide impedance matching. Figure 6.13a shows a screw coming down partly. Its effect is to increase both C_1, and C_2 and hence β increases.

To decrease β, screw or post should extend through the waveguide completely making contact with both top and bottom walls as shown in Fig. 6.13b. The effect is the same as iris in Fig. 6.12a, and hence β decreases.

Screw comes from top to bottom. Posts go bottom to up.

6.3.5 Plungers

As discussed before in tapered transmission lines, we can change a and b locally and change $\beta(z)$ by the equation

$$\beta(z) = \sqrt{\frac{\omega^2}{c^2} - \left(\frac{\pi}{a(z)}\right)^2 - \left(\frac{\pi}{b(z)}\right)^2}. \tag{6.44}$$

This can be achieved with the help of plungers as shown in Fig. 6.14a, b.

6.3.6 RF Oscillators

RF oscillators can be made with inductors and capacitors in series or parallel. Figure 6.15a shows the series version, where v_0 is the applied voltage and v_1 is the voltage across the capacitor. The current through the capacitor is $c\frac{dv_1}{dt}$, and hence the voltage across the inductor is $L\frac{d}{dt}(C\frac{dv_1}{dt})$, giving

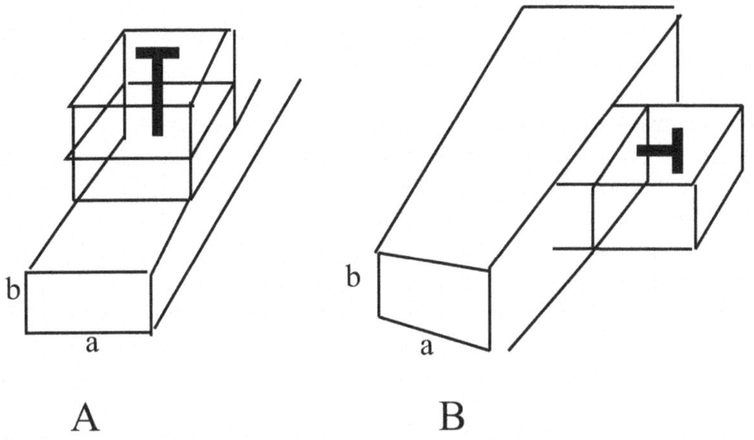

Fig. 6.14 Cross section of a waveguide changing using plungers

Fig. 6.15 (**a**) Oscillator that uses a voltage source. (**b**) Oscillator that uses a current source. (**c**) Transistor implementation of (**b**)

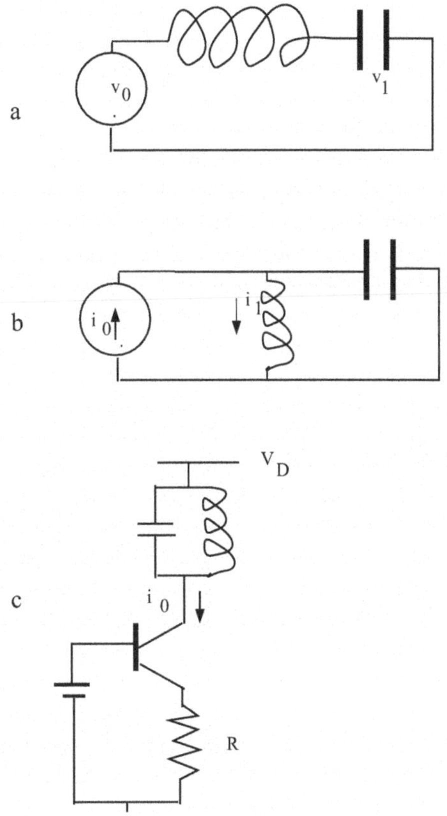

$$v_0 - v_1 = LC\frac{d^2v_1}{dt^2}, \tag{6.45}$$

with $\omega_0 = \frac{1}{\sqrt{LC}}$,

$$\omega_0^2 v_0 = \frac{d^2v_1}{dt^2} + \omega_0^2 v_1. \tag{6.46}$$

Let $x = v_1$ and $y = \frac{1}{\omega_0}\frac{dv_1}{dt}$.
Then

$$\frac{d}{dt}\begin{bmatrix} x \\ y \end{bmatrix} = \begin{bmatrix} 0 & \omega_0 \\ -\omega_0 & 0 \end{bmatrix}\begin{bmatrix} x \\ y \end{bmatrix} + \begin{bmatrix} 0 \\ \omega_0 v_0 \end{bmatrix}. \tag{6.47}$$

Using the *variation of constant formula*,

$$\begin{bmatrix} x(t) \\ y(t) \end{bmatrix} = \omega_0 v_0 \int_0^t \begin{bmatrix} \sin\omega_0\sigma \\ \cos\omega_0\sigma \end{bmatrix} d\sigma = v_0 \begin{bmatrix} 1 - \cos\omega_0 t \\ \sin\omega_0 t \end{bmatrix}. \tag{6.48}$$

We find the voltage v_1 is oscillating.

Figure 6.15b shows the parallel version of RF oscillator where we use a current source. Figure 6.15c shows the implementation of a current source using a transistor circuit. The current through inductor is i_2, and therefore voltage across it is $L\frac{di_1}{dt}$, which is also the voltage across capacitor, and hence the current through capacitor is $C\frac{d}{dt}(L\frac{di_1}{dt})$, which gives

$$i_0 - i_1 = LC\frac{d^2i_1}{dt^2}, \tag{6.49}$$

with $\omega_0 = \frac{1}{\sqrt{LC}}$,

$$\omega_0^2 i_0 = \frac{d^2i_1}{dt^2} + \omega_0^2 i_1. \tag{6.50}$$

Let $x = i_1$ and $y = \frac{1}{\omega_0}\frac{di_1}{dt}$.
Then

$$\frac{d}{dt}\begin{bmatrix} x \\ y \end{bmatrix} = \begin{bmatrix} 0 & \omega_0 \\ -\omega_0 & 0 \end{bmatrix}\begin{bmatrix} x \\ y \end{bmatrix} + \begin{bmatrix} 0 \\ \omega_0 i_0 \end{bmatrix}. \tag{6.51}$$

Using the *variation of constant formula*,

$$\begin{bmatrix} x(t) \\ y(t) \end{bmatrix} = \omega_0 i_0 \int_0^t \begin{bmatrix} \sin \omega_0 \sigma \\ \cos \omega_0 \sigma \end{bmatrix} d\sigma = i_0 \begin{bmatrix} 1 - \cos \omega_0 t \\ \sin \omega_0 t \end{bmatrix}. \tag{6.52}$$

We find the voltage across inductor y is oscillating.

Of course, oscillators have resistance, so we need some positive feedback to overcome the loss. In addition, we need to rectify the positive feedback to have a signal dependent gain. Some common oscillators based on this principle are Collpitt oscillator and Hartley oscillator.

6.4 Tank Circuits

Figure 6.16 shows a tank circuit to amplify small oscillating voltage $v_0 = A \cos \omega_0 t$. Let v_1 be the voltage across the capacitor. The current through the capacitor is $c\frac{dv_1}{dt}$, and hence the voltage across the inductor is $L\frac{d}{dt}(C\frac{dv_1}{dt})$, giving

$$v_0 - v_1 = LC\frac{d^2 v_1}{dt^2}, \tag{6.53}$$

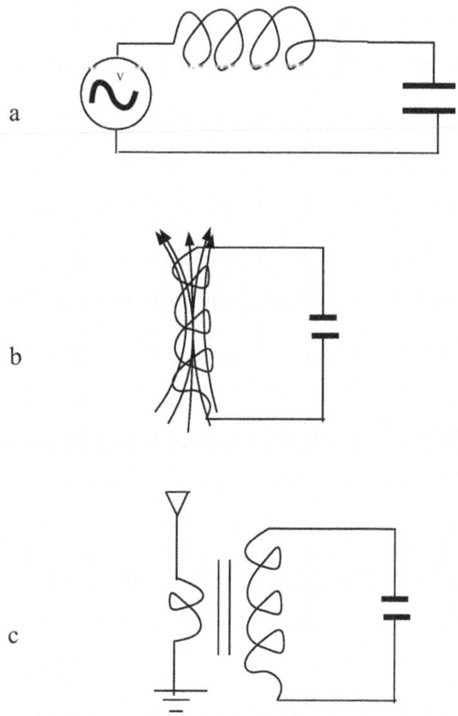

Fig. 6.16 (**a**) Tank circuit. (**b**) Tank circuit in NMR detection. (**c**) Tank circuit in a radio receiver

with $\omega_0 = \frac{1}{\sqrt{LC}}$,

$$\omega_0^2 v_0 = \frac{d^2 v_1}{dt^2} + \omega_0^2 v_1. \tag{6.54}$$

Let $x = v_1$ and $y = \frac{1}{\omega_0}\frac{dv_1}{dt}$.

Then

$$\frac{d}{dt}\begin{bmatrix} x \\ y \end{bmatrix} = \begin{bmatrix} 0 & \omega_0 \\ -\omega_0 & 0 \end{bmatrix}\begin{bmatrix} x \\ y \end{bmatrix} + \begin{bmatrix} 0 \\ \omega_0 v_0 \end{bmatrix}. \tag{6.55}$$

Using the *variation of constant formula*,

$$\begin{bmatrix} x(t) \\ y(t) \end{bmatrix} = \omega_0 A \int_0^t \cos \omega_0 (t+\sigma) \begin{bmatrix} \sin \omega_0 \sigma \\ \cos \omega_0 \sigma \end{bmatrix} d\sigma \tag{6.56}$$

$$= \frac{A\omega_0 t}{2}\begin{bmatrix} -\sin \omega_0 t \\ \cos \omega_0 t \end{bmatrix}. \tag{6.57}$$

We find that the voltage v_1 grows linearly, i.e., amplified.

6.4.1 NMR and Radio Receiver and Tank Circuit

Tank circuits are used in the field of nuclear magnetic resonance (NMR) to detect nuclear spins. The magnetic moments of nuclear spins constitute a flux passing through an inductor coil as shown in Fig. 6.16b. The precession of the spins in a magnetic field cause this flux to sinusoidally oscillate which induces a very small EMF in the coil. This can then be amplified using a tank circuit.

Tank circuits find use in radio receiver circuits as shown in Fig. 6.16c where antenna voltage is amplified using a tank circuit [14].

6.4.2 Resonant Radio Antenna and Tank Circuit

A Hertz dipole antenna is shown in Fig. 6.17a. The applied voltage pulls charges toward it, making ends of the antenna charged. This charge opposes the pull. If we bend the ends as shown in Fig. 6.17b, it begins to look like a capacitor, where charges on capacitor plate oppose the pull. Therefore, the antenna has a capacitance and has inductance as current flowing in antenna produces a magnetic field, and change in this current or resulting magnetic field produces an EMF that opposes the change, and finally the antenna has some resistance. Therefore, antenna is like an

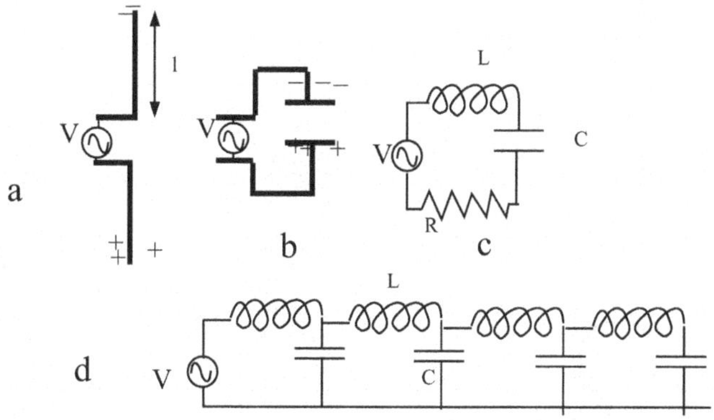

Fig. 6.17 Resonant antenna acting as a tank circuit and having gain inbuilt in it

resistance, inductance, capacitance (RLC) circuit as shown in Fig. 6.17c. Neglecting resistance for a moment, the resonant frequency of this circuit is $\omega_0 = \frac{1}{\sqrt{LC}}$. If driving voltage V is at this frequency, the impedance of the antenna vanishes, and we see a large current flow in it and antenna efficiently transmits. Of course, in Fig. 6.17a, charge is not just concentrated at the ends, but there is a charge density along the length. Hence, the capacitance is distributed in the form of a transmission line as shown in Fig. 6.17d. If we choose the length of line (antenna) as l such that $\beta l = \frac{\pi}{2}$, then the impedance of the antenna is $-jZ_0 \cot \beta l$ which becomes 0 at $\beta l = \frac{\pi}{2}$ and huge current flows (of course, we are neglecting resistance) and antenna transmits efficiently.

We discussed antenna in the transmit mode. In the receiver mode, the incident electric field acts as a voltage source, and the received voltage is the voltage across the capacitor as shown in Fig. 6.17c. As shown in our discussion of tank circuits, once the driving frequency is on resonance, this voltage builds like crazy and is only limited by the resistance of the circuit. Once again, we can be more realistic and treat antenna as a transmission line. We excite it at one end with voltage V_0 and find the open circuit voltage at the other end. Since there is no current at the open end, $V_+ = V_-$, and voltage at the driving end is

$$V_0 = V_+ \exp(-j\beta l) + V_- \exp(j\beta l) = V_+ \exp(-j\beta l)(1 + \exp(j2\beta l)). \qquad (6.58)$$

When $2\beta l = \pi$, we get $V_+ \to \infty$. Thus, transmission line amplifies the driving voltage V_0. Of course, there is a resistance, and infinity only means large gain as in a tank circuit.

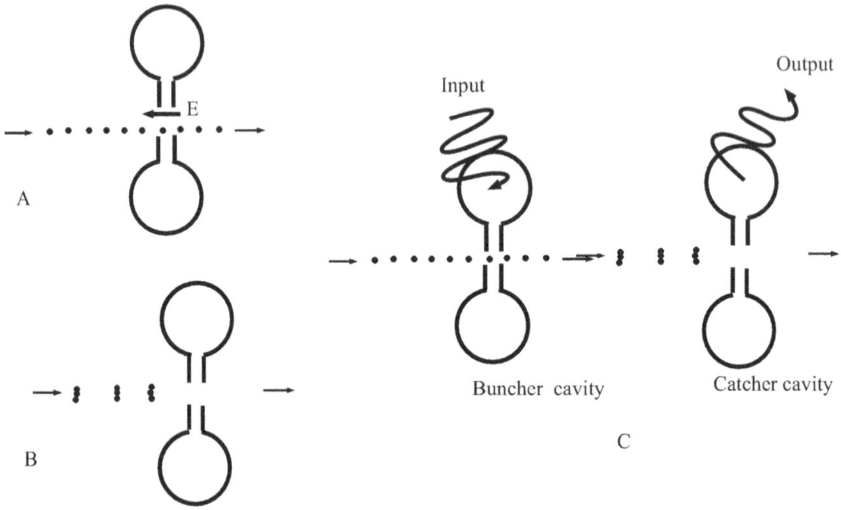

Fig. 6.18 Schematic of a microwave amplifier called klystron

6.5 Klystron Microwave Amplifier

Consider a microwave cavity as in Fig. 6.18a. Suppose we pass electrons through it as shown. These will excite cavity mode as discussed in the klystron oscillator. The electrons come one by one and trigger cavity excitation at different times. These modes are not in sync and cavity excitation does not build up. But suppose we can bunch these electrons as shown in Fig. 6.18b so that excitation times are in resonance with the cavity, then the excitations due to electron will coherently add, and we will see cavity fill up with microwaves. Now how do we bunch these electrons? We pass the electrons through an input cavity containing input microwaves that we want to amplify. The electric field due to these input signal will oscillate and oppose and accelerate the incoming electrons in a cycle. The result is in a cycle; it ends up bunching them. The bunching is more if the input signal is more, i.e., electric field is stronger. Then these bunched electrons go and fill the output cavity. When bunching is stronger, the output is stronger. Thus, we get a linear amplifier called klystron microwave amplifier.

6.6 Transistor RF Amplifiers

RF signals can be amplified by a solid-state device called transistor [30]. A transistor is made of semiconductor material like silicon. Silicon has four electrons in its outer shell. In a solid crystal, neighboring silicon atoms interact, and these electrons of the silicon lattice organize themselves into wavestates whose energies lie in a band.

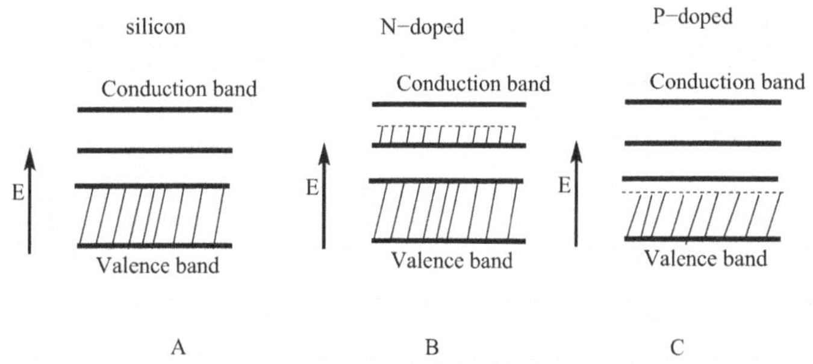

Fig. 6.19 Electrons in conduction and valence band of a doped and undoped semiconductor silicon

Wavestates are indexed by their wavenumber, and each wavestate can carry two electrons. When all wavestates in a band are full as in silicon, we cannot conduct as conduction means increasing momentum or wavenumber which is not possible as the band is full. When we replace some of the silicon atoms with say phosphorus, then we get extra electrons (phosphorus has five valence electrons instead of four) which go into next available energy band called conduction band which is not full and can conduct. We say we have *n-doped* silicon. Similarly, when you replace some of the silicon atoms with say aluminum (aluminum has three valence electrons instead of four), then we get less electrons than native silicon which means some of wavestates of the filled band (also called valence band) become available, and it can conduct. Valence band has lower energy than the conduction band. The energy gap is called the band gap. Figure 6.19a shows bands for silicon, b shows bands for n-doped silicon, and C shows bands for p-doped silicon.

When we bring an n-doped silicon next to a p-doped silicon, we form a junction called a pn junction. Some conduction electrons will flow from the n-region go the valence electrons in p-region to minimize energy. This will result in the buildup of $+$ and $-$ charge in the n and p-region at the junction, which creates an electric field that eventually halts this flow. This interface filed corresponds to a voltage of 0.7 V. When we apply a positive voltage to the p-region compared to n-region, we pull valence electrons from p-region which are substituted by the flow of conduction electrons from n-region and current flows. For this to happen, the applied voltage should be larger to overcome the interface field which is roughly 0.7 V. We say we have forward biased the pn junction. If instead we apply positive pull on n electrons, we cannot pull anything as the valence electrons in n-region are immobile, and the conduction electrons can only be substituted by valence electrons from p-region, but they have much lower energy. Hence, current cannot flow and we say we have reverse biased the pn junction, which only conducts in one direction.

Transistor has two junctions as shown in Fig. 6.20a. A p-region called base sandwiched between two n-regions (called emitter and collector) as shown in Fig. 6.20a. One pn junction is forward biased (emitter and base) and other reverse biased (base and collector). Forward-biased pn junctions suck valence electrons

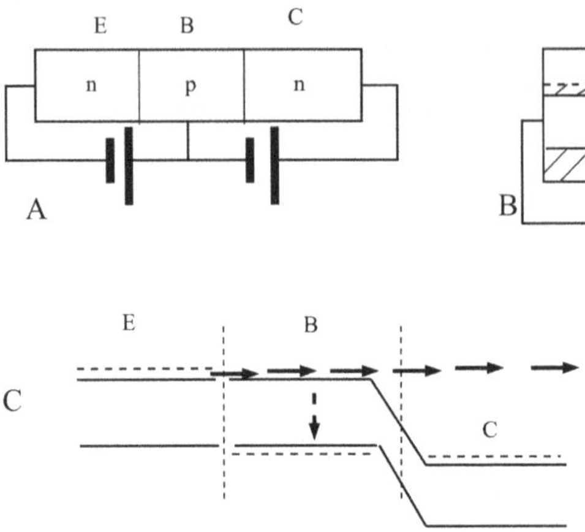

Fig. 6.20 (**a**) Schematic of a transistor. (**b**) Electrons and bands in various regions without taking into account interface electric fields. (**c**) Energy level diagram of a transistor

from p-region and **hurls** electrons from the n-region in the p-region as shown in Fig. 6.20b. These electron are immediately sucked by the reverse-biased junction and on their fraction of them α recombine and contribute to the p junction current the remaining $1 - \alpha$ fraction just fly by to other n-region as shown in Fig. 6.20b. If I is the base current, $\frac{1-\alpha}{\alpha} I = \beta I$ is the collector current and $(1 + \beta)I$ the emitter current. α is small so β is large and is called the current gain of the transistor. Figure 6.20c shows the energy level diagram for the transistor.

6.6.1 Current Source and Amplifier

Transistor acts as a current source. Figure 6.21a shows the use of a forward bias and resistance R_1; we establish a current in emitter, simply given by $I_e = \frac{V_1 - .7}{R_1}$. This gives a collector current $\frac{\beta}{\beta+1} I_e \sim I_e$ and the potential drop across resistance R_2 as $I_e R_2 = \frac{R_2(V_1 - .7)}{R_1}$. Thus, the voltage V_1 is amplified to $\frac{R_2}{R_1} V_1$, where $\frac{R_2}{R_1}$ is the amplifier gain. V_1 can have oscillating parts as shown in Fig. 6.21b, which get amplified; the DC part of V_1 is used to bias the base emitter circuit as shown in Fig. 6.21b.

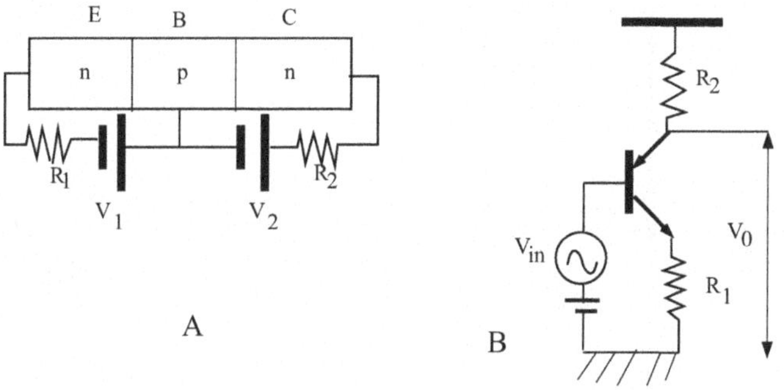

Fig. 6.21 (**a**) Schematic of a transistor. (**b**) Electrons and bands in various regions without taking into account interface electric fields. (**c**) Energy level diagram of a transistor

6.7 RF and Microwave Filters

We discussed RF and microwave oscillators. Q of the oscillator limits the purity of resulting frequency source. The source signal may contain higher harmonics. We can improve signal purity by passing it through a band pass filter. We mention here cavity filters. We discussed cavity with transfer function

$$G(s) = \frac{1}{s - j\omega_0 + \alpha^2} \tag{6.59}$$

The factor $Q = \frac{\omega_0}{\alpha^2}$ measures the quality factor of the band-pass filter which can be high as 10^6. The input port to the cavity couples the input signal. The signal at cavity resonance has large gain while other frequencies get attenuated. Filter also find natural application when a signal with noise needs to be amplified; we first filter out noise.

6.8 Cavity Coupling

The power between the cavity resonator and signal source or load can be coupled by means of a coaxial line, whose center conductor is extended inside the cavity in the form of a probe Fig. 6.22a or loop Fig. 6.22b. The cavity may be coupled to a waveguide by means of an aperture or a slot in the main wall as shown in Fig. 6.22c, d [7].

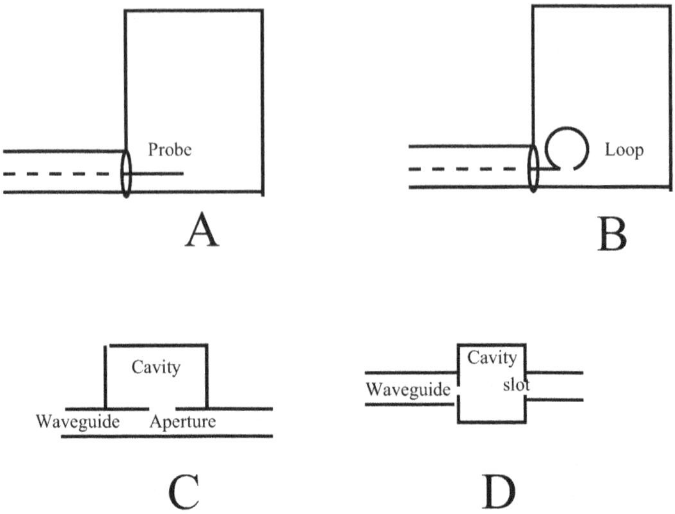

Fig. 6.22 (**a**) Cavity coupled to a coax capacitively (probe coupling). (**b**) Shows a cavity coupled to a coax inductively (loop coupling). (**c**) Shows waveguide/cavity coupled through aperture. (**d**) Shows waveguide/cavity coupled through slot

Chapter 7
Geometric Optics

7.1 Pinhole

We can model our eye as a pinhole camera, as shown in Fig. 7.1a. The light rays from object, after passing through pinhole form an image on the screen as shown in Fig. 7.1a. This does not let very much light in. How can we gather all the light coming from the object? We can increase the size of the pinhole as shown in Fig. 7.1b. This produces a very diffuse image as light from a point is spread all over the screen as shown in Fig. 7.1b. How can we collect more light coming from the object and yet have a sharp image? We need a *lens* as shown in Fig. 7.1c [13]. The light emanating from each point on the object gets focused as shown in Fig. 7.1c. How does the lens work?

7.2 Lens

Figure 7.2 shows a dense spherical medium like glass; r is the radius of the sphere. Light rays parallel to equator move into a rare medium like air. Consider a light ray making an angle α with radial line as shown in Fig. 7.2.

From Snell's laws of refraction,

$$\frac{\sin \alpha + \beta}{\sin \alpha} = C. \tag{7.1}$$

$$r \sin \alpha = d \tan \beta, \tag{7.2}$$

for small α, β, we have

N. Khaneja, *Electromagnetic Waves and Optics*,
https://doi.org/10.1007/978-3-031-76144-7_7

Fig. 7.1 (**a**) Pinhole camera. (**b**) Diffuse image if we increase the size of pinhole. (**c**) Light focused with a lens

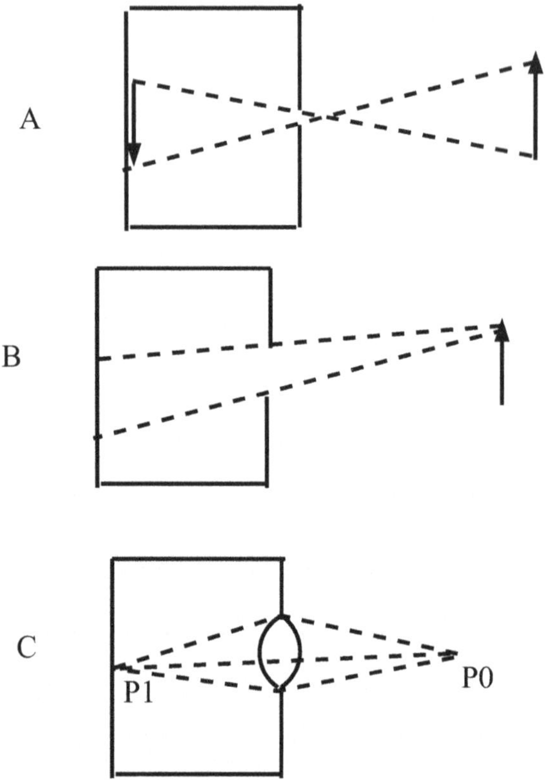

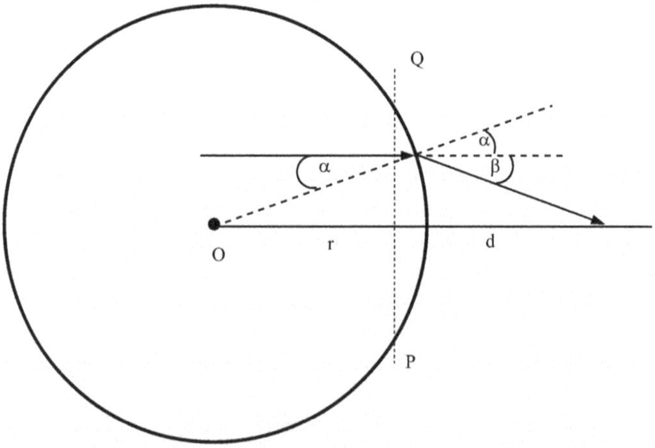

Fig. 7.2 Light moving from dense spherical medium to rare medium

$$\frac{\alpha + \beta}{\alpha} = C, \tag{7.3}$$

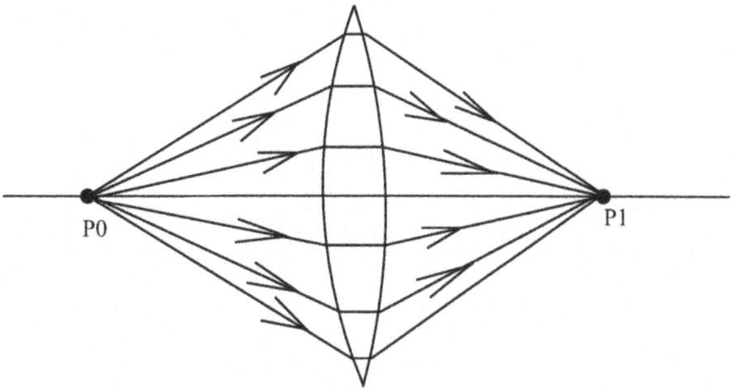

Fig. 7.3 Light rays emanating from 2F point becoming parallel inside the lens and focusing on the 2F point on the other side

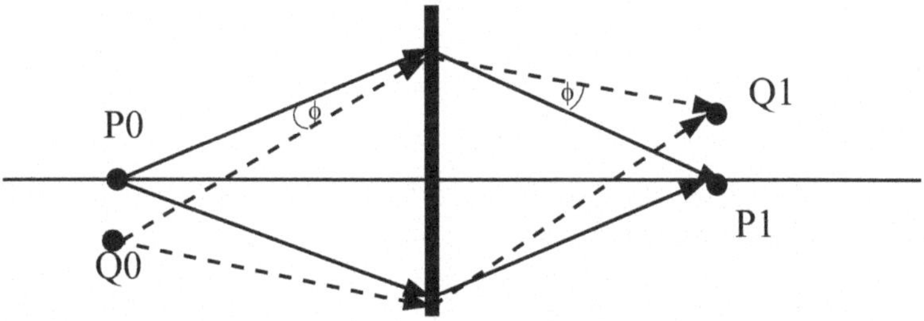

Fig. 7.4 Light from point Q_0 shifted from 2F point converging to Q_1 on the other side and is shifted the other way

$$\frac{\alpha}{\beta} = (C - 1)^{-1} = C'. \tag{7.4}$$

$$d = \frac{r}{C'}. \tag{7.5}$$

Irrespective of α, we focus at same d as above. We call $\frac{d}{2} = f$ the focal point and f focal length. We call d the 2F point. If we reverse the light direction, we will start from the focal point and when light rays enter the glass medium, they become parallel. We assume all action is near equator so that α, β are small. If we cut this sphere at the periphery, as shown by dotted line QP in Fig. 7.2, we get a half lens, and we join two such half lenses back to back to form a lens as shown in Fig. 7.3. Light rays emanating from 2F point become parallel inside the lens and focus on the 2F point on the other side. The two lenses need not have the same focal length. We can have two focal lengths f_1 and f_2. This is how the eye lens works. The focal length of the half lens that sees the world is changed to focus objects at different distances, while the half lens facing the retina focuses everything on the retina, and its focal length is not changed that much.

We say how light rays emerging from a 2F point become parallel and focused again on the 2F point on the other side of lens. How about a point that is displaced vertically from the 2F point? The light rays tilt by an angle ϕ as shown in Fig. 7.4. Figure 7.4 shows that the point Q_0 converges to Q_1 on the other side that is shifted

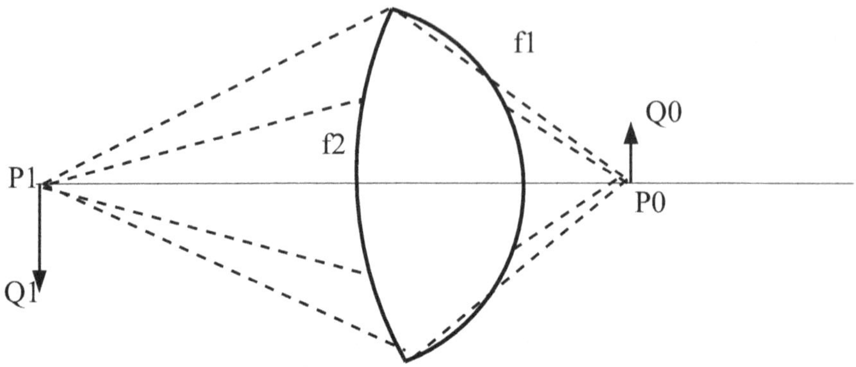

Fig. 7.5 Magnification principle of a microscope with $P0Q0$ magnified to $P1Q1$

the other way. If focal lengths are different and if Q_0 is displaced by x, Q_1 displaced by y; then for ϕ as shown in Fig. 7.4,

$$\phi = \frac{x}{f_1} = \frac{y}{f_2}, \tag{7.6}$$

which gives Q_1 is displaced by $\frac{f_2 x}{f_1}$.

7.3 Microscopes

This forms the basis of what is called a microscope. If $f_2 \gg f_1$, the object gets magnified. This is shown in Fig. 7.5.

7.4 Telescopes

Light coming from a distant object can be first focused with a lens (objective lens) and then magnified by another lens (eye piece) as in microscope to form what is called a telescope that helps us see far. This is shown in Fig. 7.6. With bare eyes, the image on retina is negligibly small. In telescope, the microscope action first enlarges the image $P1$, before it is seen.

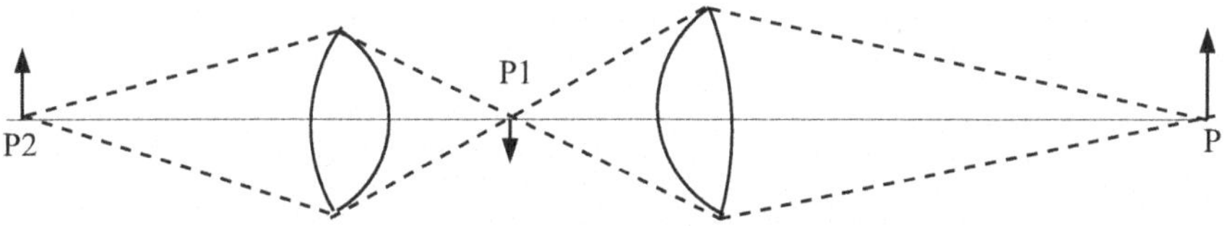

Fig. 7.6 Magnification principle of a telescope, with P magnified to $P2$

7.5 Resolution and Diffraction

Microscopes can magnify. Is there any limit? Can we magnify arbitrary small objects? We find that the small size is limited by the wavelength of light used. Consider a small slit in the wall in Fig. 7.7 of aperture size D. Light coming from the left passes through the slit and goes to the right but not in straight line. The slit acts as an antenna array and light spreads out, a phenomenon called diffraction. The spread θ of light in Fig. 7.7 is as described in the treatment of antenna arrays and given by

$$kD\theta = \pi, \tag{7.7}$$

where $k = \frac{2\pi}{\lambda}$ or $\theta = \frac{\lambda}{2D}$. To resolve O and P, we have to make sure $\theta < \alpha$ or

$$\frac{\lambda}{2D} < \frac{x}{2f}, \tag{7.8}$$

$$x > \frac{f\lambda}{D}. \tag{7.9}$$

Fig. 7.7 Diffraction limiting resolution of magnification

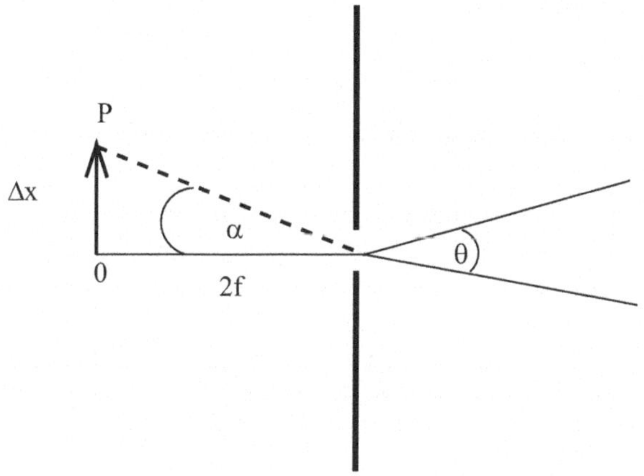

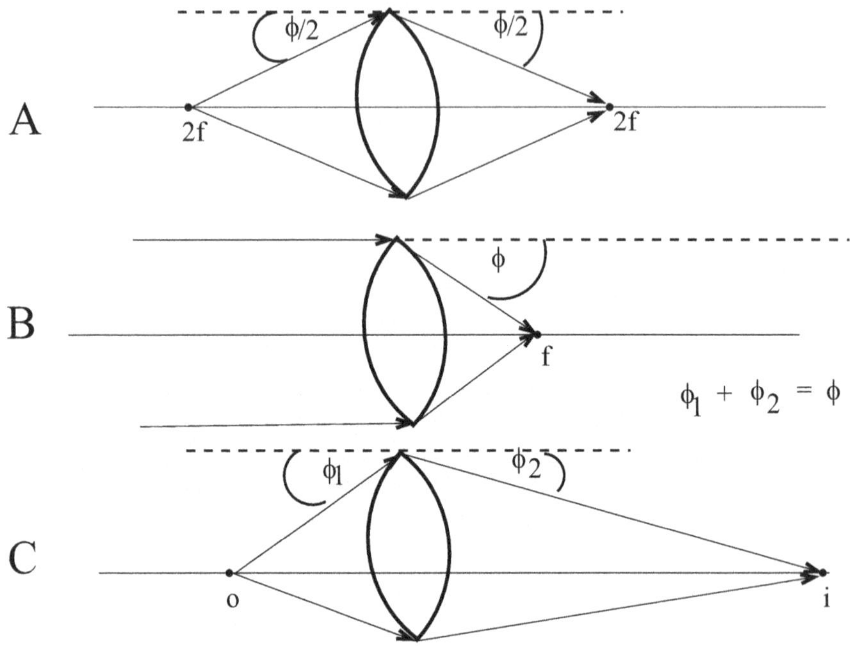

Fig. 7.8 (**a**) For object at $2f$, light rays converging on the other side at $2f$. They turn around by angle 2ϕ. (**b**) Parallel light rays passing through focal point. (**c**) Object and image distances related by lens equation

The limiting size is $x = \frac{f\lambda}{D}$, which becomes smaller if we use smaller wavelength light or larger D. We can now simply replace slit with a lens with D as aperture of the lens, and the above analysis shows how small an object can be for it to be magnified.

7.6 Lens Equation

Until now, we showed the design of optical instruments with lenses whose half lenses were of different focal lengths. However, it is possible to do everything with lenses with half lenses of the same focal length. When we put object at $2f$, light rays converge on the other side at $2f$. They turn around by angle 2ϕ, where $\phi = \frac{x}{2f}$, where x is vertical distance from lens center as shown in Fig. 7.8a. If light rays started parallel when they turn around by 2ϕ, they will pass through the focal point as shown in Fig. 7.8b. In general, when they emanate from point o as shown in Fig. 7.8c, the angle $\phi_1 = \frac{x}{o}$, and when they converge at point i as shown in Fig. 7.8c, the angle $\phi_2 = \frac{x}{i}$. They turn by $\phi + \phi_2 = 2\phi$, which gives the famous lens equation

$$\frac{1}{i} + \frac{1}{o} = \frac{1}{f}. \tag{7.10}$$

7.7 Revisiting Optical Instruments

Now we revisit the design of optical instruments using lens equation. When we move the object to be magnified close to f, the image point i moves far and object size x gets magnified to $\frac{ix}{o}$. This is shown in Fig. 7.9. This forms the basis of a microscope.

Figure 7.10 shows the design of telescope using lens equation. The object is further than $2f$ for the objective lens whose image is formed between f and $2f$ of the objective. This image is between f and $2f$ of the eye piece and is then magnified.

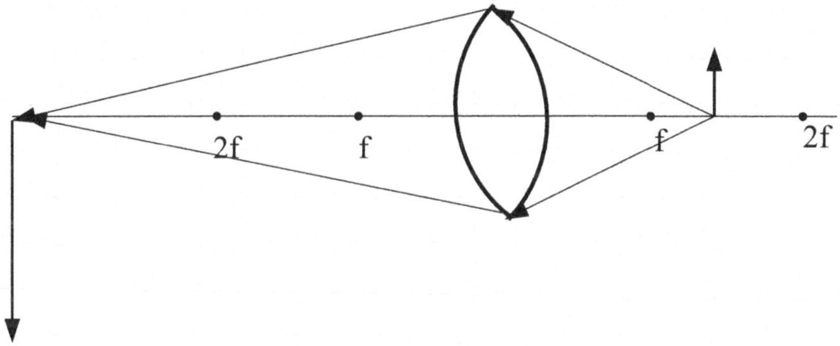

Fig. 7.9 Design of a microscope using lens equation

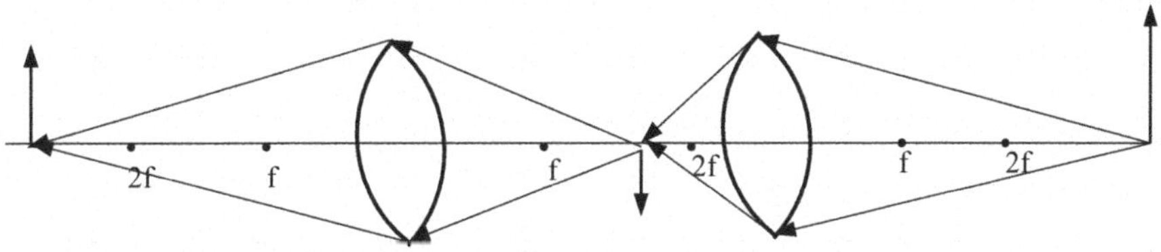

Fig. 7.10 Design of a telescope using lens equation

7.8 Problems

1. If object distance to a lens of focal length 10 cm is 20 cm, what is the image distance?
2. In a microscope, object distance to objective lens of focal length 10 cm is 20 cm. If the distance between objective and eye piece (focal length 10 cm) is 35 cm, what is the image magnification?
3. In a telescope, the distance between objective and eye piece is 22.5 cm. If both lenses have a focal length of 10 cm, what is the magnification of telescope?
4. The size of a pinhole camera is 5 cm. What is the image size if the object is 5 m from the camera and 1 m tall.
5. For a light of 500 nm wavelength, what is the smallest size you can resolve, with an aperture of 5 cm.

Chapter 8
Optical Resonance and Colors

8.1 Light Absorption by an Atom

Atom has electrons organized in orbitals. An electron from higher energy state (excited orbital) can jump to a lower energy state (ground-state orbital) giving radiation with frequency ω_0, where $\hbar\omega_0$ is the energy difference between the two states ($\hbar$ is the Planck's constant). Similarly, when light is shone on an atom, with frequency ω_0, it can absorb light and jump from lower energy state to higher energy state.

Consider an EM wave incident on an atom with electron in ground state. We take the EM field as plane wave traveling along x direction and electric field $E_0 \exp(j\omega_0 t)\hat{z}$. Let us denote ground-state orbital by ϕ_0 and the excited state by ϕ_1. Electron can be in the superposition $a\phi_0 + b\phi_1$. In this superposition state, it has a dipole moment along x axis as

$$-e\langle a\phi_0 + b\phi_1 | x | a\phi_0 + b\phi_1 \rangle = d(ab^* + a^* b), \tag{8.1}$$

where $d = -e\langle \phi_1 | x | \phi_0 \rangle = -e \int \phi_1^* \phi_0 x$, which we take real.

When an electric field is applied say in the x direction, it produces a potential

$$V(x) = -eEx \tag{8.2}$$

at the atomic site. This potential scatters ϕ_0 to ϕ, and we have dipole matrix element

$$d = -e\langle \phi_1 | x | \phi_0 \rangle. \tag{8.3}$$

Let the state of the system be given by $\psi = a(t)\phi_0 + b(t)\phi_1$ which we write as $\psi(t) = \begin{bmatrix} a(t) \\ b(t) \end{bmatrix}$. Then if $h\omega_0$ is the difference in energies of ϕ_0 and ϕ_1 and $E = E_0 \exp(j\omega_0 t)$, then $\psi(t)$ evolves as

N. Khaneja, *Electromagnetic Waves and Optics*,
https://doi.org/10.1007/978-3-031-76144-7_8

$$\dot{\psi} = -\frac{i}{2}\begin{bmatrix} \omega_0 & \frac{d^* E_0}{\hbar}\exp(-j\omega_0 t) \\ \frac{dE_0}{\hbar}\exp(j\omega_0 t) & -\omega_0 \end{bmatrix}\psi. \tag{8.4}$$

Let $\frac{dE_0}{\hbar} = \omega_1$,

$$\Psi(t) = \exp\left(\frac{i}{2}\begin{bmatrix} \omega_0 & 0 \\ & -\omega_0 \end{bmatrix} t\right)\Psi(t), \tag{8.5}$$

then

$$\dot{\Psi} = -\frac{i}{2}\begin{bmatrix} 0 & \omega_1 \\ \omega_1 & 0 \end{bmatrix}\Psi, \tag{8.6}$$

then

$$\Psi(0) = \begin{bmatrix} 1 \\ 0 \end{bmatrix} \rightarrow \begin{bmatrix} \cos\left(\frac{\omega_1 t}{2}\right) \\ -j\sin\left(\frac{\omega_1 t}{2}\right) \end{bmatrix}. \tag{8.7}$$

$$\psi(0) = \begin{bmatrix} 1 \\ 0 \end{bmatrix} \rightarrow \begin{bmatrix} \exp\left(-j\frac{\omega_0 t}{2}\right)\cos\left(\frac{\omega_1 t}{2}\right) \\ -j\exp\left(j\frac{\omega_0 t}{2}\right)\sin\left(\frac{\omega_1 t}{2}\right) \end{bmatrix}. \tag{8.8}$$

At time $\omega_1 T = \pi$, we go to the excited state. This is resonant absorption of light.

8.2 Light Scattering by an Atom

Input light need not be resonant to optical transition and can be far detuned, in which case the light photon cannot be absorbed as we do not conserve energy. The energy of the photon $\hbar\omega$ and ground atomic state is not equal to the energy of the excited state. Instead what can happen is we go to the intermediate excited state and then return to the ground state and emit photon may be in a different direction. We say we have scattered. Scattering is understood as a three-level atomic system, the so-called Λ system. In a Λ system as shown in Fig. 8.1 we have two ground-state levels $|1\rangle$ (ground-state atom and input photon) and $|3\rangle$ (ground-state atom and output photon) at energy E_1 and excited level (excited atom) $|2\rangle$ at energy E_2. The transition from $|1\rangle$ to $|2\rangle$ has strength Ω_1 and transition from $|2\rangle$ to $|3\rangle$ has strength Ω_2. In the interaction frame of natural Hamiltonian of the system, we get a second-order term connecting level $|1\rangle$ to $|3\rangle$ with strength $\frac{\Omega_1\Omega_2}{(E_1-E_2)}$. This term creates an effective coupling between ground-state levels and drives transition from $|1\rangle$ to $|3\rangle$. To see this, observe

The state of the three-level system evolves according to the Schröedinger equation

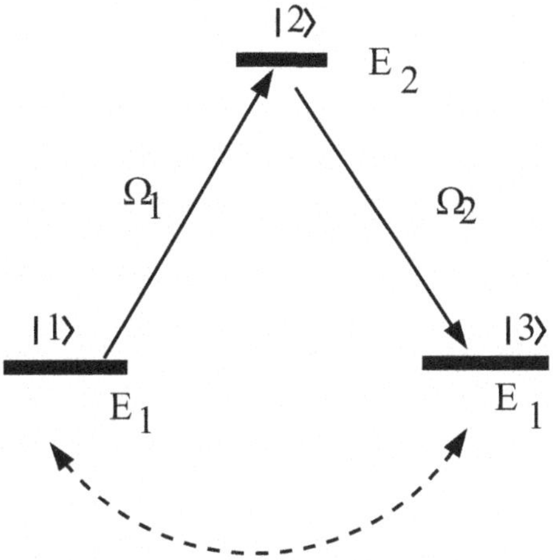

Fig. 8.1 A three-level Λ system with two ground-state levels |1⟩ and |3⟩ and an excited level |2⟩

$$\dot{\psi} = \frac{-i}{\hbar} \begin{bmatrix} E_1 & \Omega_1^* & 0 \\ \Omega_1 & E_2 & \Omega_2^* \\ 0 & \Omega_2 & E_1 \end{bmatrix} \psi. \tag{8.9}$$

We proceed into the interaction frame of the natural Hamiltonian (system energies) by transformation

$$\phi = \exp\left(\frac{i}{\hbar} \begin{bmatrix} E_1 & 0 & 0 \\ 0 & E_2 & 0 \\ 0 & 0 & E_1 \end{bmatrix} \right) \psi. \tag{8.10}$$

This gives for $\Delta E = E_2 - E_1$,

$$\dot{\phi} = \underbrace{\frac{-i}{\hbar} \begin{bmatrix} 0 & \exp(-\frac{i}{\hbar}\Delta E\, t)\Omega_1^* & 0 \\ \exp(\frac{i}{\hbar}\Delta E\, t)\Omega_1 & 0 & \exp(\frac{i}{\hbar}\Delta E\, t)\Omega_2^* \\ 0 & \exp(-\frac{i}{\hbar}\Delta E\, t)\Omega_2 & 0 \end{bmatrix}}_{H(t)} \phi. \tag{8.11}$$

$H(t)$ is periodic with period $\Delta t = \frac{2\pi}{\Delta E}$. After Δt, the system evolution is

$$\phi(\Delta t) = \left(I + \int_0^{\Delta t} H(\sigma) d\sigma + \int_0^{\Delta t} \int_0^{\sigma_1} H(\sigma_1) H(\sigma_2) d\sigma_2 d\sigma_1 + \ldots \right) \phi(0). \tag{8.12}$$

The first integral averages to zero, while the second integral

$$\int_0^{\Delta t} \int_0^{\sigma_1} H(\sigma_1) H(\sigma_2) d\sigma_2 d\sigma_1 = \frac{1}{2} \int_0^{\Delta t} \int_0^{\sigma_1} [H(\sigma_1), H(\sigma_2)] d\sigma_2 d\sigma_1. \tag{8.13}$$

Evaluating it explicitly, we get for our system that second-order integral is

$$\frac{-i\Delta t}{\hbar}\begin{bmatrix} 0 & & \frac{\Omega_1^*\Omega_2^*}{E_1-E_2} \\ 0 & 0 & 0 \\ \underbrace{\frac{\Omega_1\Omega_2}{E_1-E_2}}_{\mathcal{M}} & 0 & 0 \end{bmatrix}. \tag{8.14}$$

Thus, we have created an effective Hamiltonian

$$\begin{bmatrix} 0 & & \frac{\Omega_1^*\Omega_2^*}{E_1-E_2} \\ 0 & 0 & 0 \\ \frac{\Omega_1\Omega_2}{E_1-E_2} & 0 & 0 \end{bmatrix}, \tag{8.15}$$

which couples levels $|1\rangle$ and $|3\rangle$ and drives transition between them at rate

$$\mathcal{M} = \frac{\Omega_1\Omega_2}{(E_1 - E_2)} \tag{8.16}$$

So in scattering, we scatter input photon to a different direction output photon.

8.2.1 Electric Dipoles

Recall Maxwell equation in a medium

$$\nabla \times E = -\frac{\partial B}{\partial t}, \tag{8.17}$$

$$\nabla \times H = \frac{\partial D}{\partial t}. \tag{8.18}$$

$$\nabla \times E = -j\mu\omega H, \tag{8.19}$$

$$\nabla \times H = j\epsilon\omega E. \tag{8.20}$$

In a material medium, we have

$$\epsilon = \epsilon_0(1 + \chi(\omega)), \tag{8.21}$$

$$\chi(\omega) = \chi'(\omega) + j\chi''(\omega). \tag{8.22}$$

$\chi(\omega)$ is called optical susceptibility and depends on the frequency. It has a real part $\chi'(\omega)$ and an imaginary part $\chi''(\omega)$.

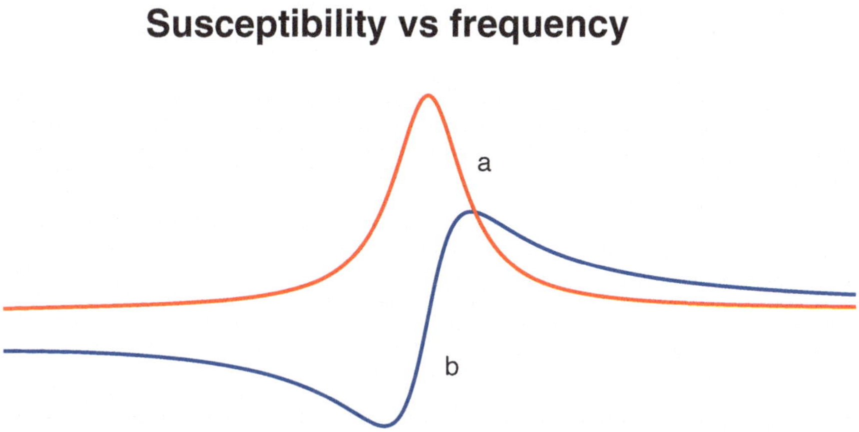

Fig. 8.2 This figure shows real (**a**) and imaginary (**b**) part of susceptibility as function of frequency

$$\sqrt{\frac{\epsilon}{\epsilon_0}} = \sqrt{1 + \chi'(\omega) + j\chi''(\omega)} = n + j\alpha, \tag{8.23}$$

n is the index of refraction and α is the absorption coefficient. Let us understand the source of $\chi(\omega)$, the optical susceptibility.

8.2.2 Classical Theory of Optical Susceptibility

Electrons are bound to atoms by positive charge of the nucleus. Electric field pulls the electrons away, and we can describe the displacement as

$$m\ddot{x} + \gamma\dot{x} + kx = -eE. \tag{8.24}$$

$$\ddot{x} + \frac{\gamma}{m}\dot{x} + \omega_0^2 x = \frac{-eE}{m}, \tag{8.25}$$

where $\omega_0^2 = \frac{k}{m}$

With $E = E_0 \exp(j\omega t)$, we get

$$x(\omega) = \frac{-eE_0}{m} \frac{1}{(\omega_0^2 - \omega^2) + j\omega\frac{\gamma}{m}} \tag{8.26}$$

Susceptibility is proportional to x. Figure 8.2a, b gives the imaginary and real part of susceptibility as function of ω. The imaginary part peaks as $\omega = \omega_0$.

8.2.3 Quantum Theory of Optical Susceptibility

Let us ϕ_0 and ϕ_1 denote two electronic states (orbitals) of an atom. Let ϕ_0 be ground state and ϕ_1 excited state. When an electric field is applied say in the x direction, it produces a potential

$$V(x) = -eEx \tag{8.27}$$

at the atomic site. This potential scatters ϕ_0 to ϕ_1, and we have dipole matrix element

$$d = -e\langle \phi_1 | x | \phi_0 \rangle. \tag{8.28}$$

Let the state of the system be given by $\psi = a(t)\phi_0 + b(t)\phi_1$, which we write as $\psi(t) = \begin{bmatrix} a(t) \\ b(t) \end{bmatrix}$. Then if $h\omega_0$ is the difference in energies of ϕ_0 and ϕ_1 and $E = 2E_0 \cos \omega t$, then $\psi(t)$ for $\omega_1 = \frac{2dE_0}{\hbar}$ evolves as

$$\dot{\psi} = -\frac{i}{2} \begin{bmatrix} \omega_0 & \omega_1 \cos(\omega t) \\ \omega_1 \cos(\omega t) & -\omega_0 \end{bmatrix} \psi. \tag{8.29}$$

Let

$$\Psi(t) = \exp\left(\frac{i}{2} \begin{bmatrix} \omega & 0 \\ & -\omega \end{bmatrix} t \right) \psi(t), \tag{8.30}$$

then

$$\dot{\Psi} = -\frac{i}{2} \begin{bmatrix} (\omega_0 - \omega) & \frac{\omega_1}{2}(1 + \exp(2j\omega t)) \\ \frac{\omega_1}{2}((1 + \exp(-2j\omega t)) & -(\omega_0 - \omega) \end{bmatrix} \Psi. \tag{8.31}$$

We neglect the fast oscillating term $\exp(2j\omega t)$ to get

$$\dot{\Psi} = -\frac{i}{2} \begin{bmatrix} (\omega_0 - \omega) & \omega_x \\ \omega_x & -(\omega_0 - \omega) \end{bmatrix} \Psi, \tag{8.32}$$

where $\omega_x = \frac{dE_0}{\hbar}$. Then we can write

$$\rho = \Psi \Psi^\dagger = \frac{\mathbf{1}}{2} + s_x \sigma_x + s_y \sigma_y + s_z \sigma_z. \tag{8.33}$$

where

$$\sigma_x = \frac{1}{2} \begin{bmatrix} 0 & 1 \\ 1 & 0 \end{bmatrix}; \ \sigma_y = \frac{1}{2} \begin{bmatrix} 0 & -i \\ i & 0 \end{bmatrix}; \ \sigma_z = \frac{1}{2} \begin{bmatrix} 1 & 0 \\ 0 & -1 \end{bmatrix}. \tag{8.34}$$

(s_x, s_y, s_x) have special meaning. s_z corresponds to population in ground and excited state. When the atom is in the ground state, $s_z = 1$ and in the excited state, $s_z = -1$. s_x, s_y correspond to real and imaginary parts of susceptibility as shown below. Then the resulting dipole is

$$\langle a\phi_0 + b\phi_1 | x | a\phi_0 + b\phi_1 \rangle, \tag{8.35}$$

$$= \left\langle \Psi_0 \exp\left(-\frac{j\omega_0 t}{2}\right)\phi_0 + \Psi_1 \exp\left(\frac{j\omega_0 t}{2}\right)\phi_1 | x | \Psi_0 \exp\left(-\frac{j\omega_0 t}{2}\right)\phi_0 \right.$$

$$\left. + \Psi_1 \exp\left(\frac{j\omega_0 t}{2}\phi_1\right)\right\rangle,$$

$$= (\Psi_1^* \Psi_0 + \Psi_1 \Psi_0^*) \cos \omega t - i(\Psi_1^* \Psi_0 - \Psi_1 \Psi_0^*) \sin \omega t, \tag{8.36}$$

$$= s_x \cos \omega t - s_y \sin \omega t. \tag{8.37}$$

From Eq. (8.32), we find that (s_x, s_y, s_z) evolves as

$$\frac{d}{dt}\begin{bmatrix} s_x \\ s_y \\ s_z \end{bmatrix} = \begin{bmatrix} 0 & \omega - \omega_0 & 0 \\ \omega_0 - \omega & 0 & -\omega_x \\ \omega_x & 0 & 0 \end{bmatrix}\begin{bmatrix} s_x \\ s_y \\ s_z \end{bmatrix}. \tag{8.38}$$

Due to dissipative processes like spontaneous emission (which we will describe soon), the above equations have two decay constants Γ_1 and Γ_2, and the above equation takes the form

$$\frac{d}{dt}\begin{bmatrix} s_x \\ s_y \\ s_z \end{bmatrix} = \begin{bmatrix} -\Gamma_2 & \Delta\omega & 0 \\ -\Delta\omega & -\Gamma_2 & -\omega_x \\ 0 & \omega_x & -\Gamma_1 \end{bmatrix}\begin{bmatrix} s_x \\ s_y \\ s_z \end{bmatrix} + \begin{bmatrix} 0 \\ 0 \\ \Gamma_1 \end{bmatrix}. \tag{8.39}$$

We can solve for the steady state of the above equations to find [29],

$$(s_x, s_y, s_z)_\infty = \frac{1}{\Gamma_2^2 + \Delta\omega^2 + \omega_x \frac{\Gamma_2}{\Gamma_1}}(\Delta\omega\, \omega_x, \omega_x \Gamma_2, \Gamma_2^2 + \Delta\omega^2). \tag{8.40}$$

Here s_x, s_y corresponds to real and the imaginary part of susceptibility, and if we plot them as a function of $\Delta\omega$, they take the form as shown in Fig. 8.2b, a, respectively. Observe imaginary part decays as a function of $\Delta\omega$. Also recall that the imaginary part is responsible for decay in a medium as in a conductor. Now optical resonance ω_0 in in energy range $\hbar\omega_0 \sim eV$ or $\omega_0 \sim 10^{14} - 10^{15}$ Hz. If we use radio waves, $\Delta\omega$ is very large, and hence $\chi''(\omega)$ is negligible, loss is negligible, and radio waves go through walls. Light does not, because $\Delta\omega$ is small, hence $\chi''(\omega)$ is not negligible, and we have attenuation or loss. Furthermore, $\chi''(\omega)$ decays much faster as a function of $\Delta\omega$ than $\chi'(\omega)$. Hence, even for radio or microwaves, we have nonzero $\chi'(\omega)$, a medium which explains dielectric response ϵ as different from ϵ_0.

Fig. 8.3 (**a**) Transition from ground to excited state. (**b**) Excited state level broadened

8.3 Spontaneous Emission

Consider a two-level system as shown in Fig. 8.3a. Let Ω be the rate of on-resonance excitation from ground state to excited state. The dynamics of the system with $\psi = \begin{bmatrix} \psi_1 \\ \psi_0 \end{bmatrix}$ is

$$\dot{\psi} = -\frac{i}{2}\begin{bmatrix} \omega_0 & \Omega\exp(-j\omega_0 t) \\ \Omega\exp(j\omega_0 t) & -\omega_0 \end{bmatrix}\psi. \tag{8.41}$$

In the rotating frame $\psi \to \exp\left(\frac{i}{2}\begin{bmatrix} \omega_0 & 0 \\ 0 & -\omega_0 \end{bmatrix} t\right)\psi$, we have

$$\dot{\psi} = -\frac{i}{2}\begin{bmatrix} 0 & \Omega \\ \Omega & 0 \end{bmatrix}\psi. \tag{8.42}$$

Now suppose the excited level is detuned by energy $\Delta\omega$, then

$$\dot{\psi} = -\frac{i}{2}\begin{bmatrix} \omega_0 + 2\Delta\omega & \Omega\exp(-j\omega_0 t) \\ \Omega\exp(j\omega_0 t) & -\omega_0 \end{bmatrix}\psi, \tag{8.43}$$

then in the rotating frame $\psi \to \exp\left(\frac{i}{2}\begin{bmatrix} \omega_0 + 2\Delta\omega & 0 \\ 0 & -\omega_0 \end{bmatrix} t\right)\psi$, we have

$$\dot{\psi} = -\frac{i}{2}\begin{bmatrix} 0 & \Omega\exp(j\Delta\omega t) \\ \Omega\exp(-j\Delta\omega t) & 0 \end{bmatrix}\psi. \tag{8.44}$$

Then starting from $\psi_0(0) = 1$, we have $\psi_1(\Delta t) = \int_0^{\Delta t} \Omega\exp(j\Delta\omega\tau)d\tau$. Now consider many excited levels as shown in Fig. 8.3b, with level ψ_n detuned by $n\Delta\omega$. Then $\psi_n(\Delta t) = \int_0^{\Delta t} \Omega\exp(jn\Delta\omega\tau)d\tau$. Then

$$\psi_k^*(\Delta t)\psi_k(\Delta t) = \int_0^{\Delta t}\int_0^{\Delta t} \Omega^2 \exp(jk\Delta\omega(t-\tau))d\tau dt. \tag{8.45}$$

Then $\Phi = \sum_{k=-n}^{n} \psi_k^*(\Delta t)\psi_k(\Delta t) =$

$$\frac{\Omega^2}{\Delta\omega}\int_0^{\Delta t}\int_0^{\Delta t}\left(\int_{-B}^{B} \exp(j\omega(t-\tau))d\omega\right) d\tau dt, \tag{8.46}$$

$$= \frac{\Omega^2}{\Delta\omega}\int_0^{\Delta t}\int_0^{\Delta t} \frac{\sin(B(t-\tau)}{t-\tau} d\tau dt. \tag{8.47}$$

$$= \frac{\Omega^2}{\Delta\omega}\Delta t. \tag{8.48}$$

where we use the approximation $B \gg \Omega$. There is time Δt such that $\Omega\Delta t \ll 1$ and $B\Delta t \gg 1$. Then

$$\Gamma = \frac{\Phi}{\Delta t} = \frac{\Omega^2}{\Delta\omega} \tag{8.49}$$

gives the transition rate out of ground state.

With Eq. (8.49), also known as *Fermi golden rule*, we can study the spontaneous emission rate of an atom. It is the rate at which atom decays into ground state. Let $\hbar\omega_0$ be the energy difference between the excited and ground state. When atom decays from excited to ground state, it radiates EM wave mode. These modes are plane waves with defined $k = (k_x, k_y, k_z)$. We give the mode some width $(\Delta k_x, \Delta k_y, \Delta k_z)$, which gives the mode a finite volume $V^{-1} = \frac{\Delta k_x \Delta k_y \Delta k_z}{(2\pi)^3}$. Figure 8.4 depicts this width for a mode with k along x axis. The energy of this mode is $\hbar\omega_0$, and hence it has the electric field $\epsilon_0 E^2 V = \hbar\omega_0$ and hence the transition rate due to this electric field $\Omega = \frac{Ed}{\hbar}$. In fact, transition rate is $\cos\theta\Omega$, where θ is the vector k vector makes with z axis as transition is induced by electric field in the equatorial plane. The mode has width Δk along radial direction and hence frequency width of $\Delta\omega = c\Delta k$. The modes are spread radially with energy spacing of $\Delta\omega$. Therefore, by Fermi golden rule, the decay rate Γ is $\Gamma = \frac{\Omega^2 \mathcal{M}}{\Delta\omega}$, where $\mathcal{M}$ is the number of modes on the surface of sphere of radius k_0. This is

$$\mathcal{M} = 2\pi k_0^2 \Delta k \int \cos^2\theta \sin\theta \, d\theta,$$

where $\cos^2\theta$ comes from the factor $\cos\theta$ in transition rate. When we put everything together, we get [24]

$$\Gamma = \frac{d^2\omega_0^3}{6\pi^2 c^3 \hbar\epsilon_0} = \frac{d^2\omega_0^3}{3\pi c^3 \hbar\epsilon_0}. \tag{8.50}$$

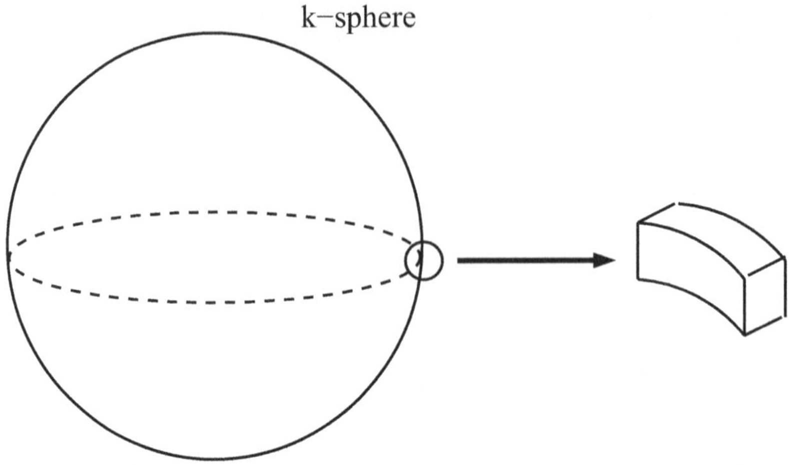

Fig. 8.4 Blowup of a mode depicting it has some width in k-space

8.3.1 *Spontaneous Emission in a Resonant Cavity and Purcell Effect*

Now consider a resonant cavity, with volume V. Let its natural frequency be ω_0. Let Q be the quality factor of the cavity and consider a two-level atom in the cavity with natural frequency ω_0. We are interested in the spontaneous emission rate of the atom in the cavity. This rate is the same rate as exciting the atom from the ground state with electric field worth $\hbar\omega_0$ worth of energy, i.e., $\epsilon_0 E_0^2 V = \hbar\omega_0$. The corresponding transition frequency is $\Omega = \frac{E_0 d}{\hbar}$. The electric field decays as $E_0 \exp(-\alpha t)$, with $\frac{\alpha}{\omega_0} = \xi$, $Q = 1 - \xi$. The transition with decaying field E can be simulated by a broadened transition with n excited levels, with frequency spacing $\Delta\omega$, such that $n\Delta\omega = \alpha$ and transition frequency $\frac{\Omega}{\sqrt{n}}$. Then by Fermi golden rule, we have

$$\Gamma = \frac{\Omega^2}{n\Delta\omega} = \frac{Q}{V}\frac{d^2}{\epsilon\hbar}. \tag{8.51}$$

Observing that $\omega_0^3 \propto \frac{c^3}{V}$, we get

$$\Gamma = Q\frac{d^2\omega_0^3}{c^3\epsilon\hbar}. \tag{8.52}$$

Thus, there is modification of spontaneous emission rate over vacuum; this modification Q is called *Purcell effect*.

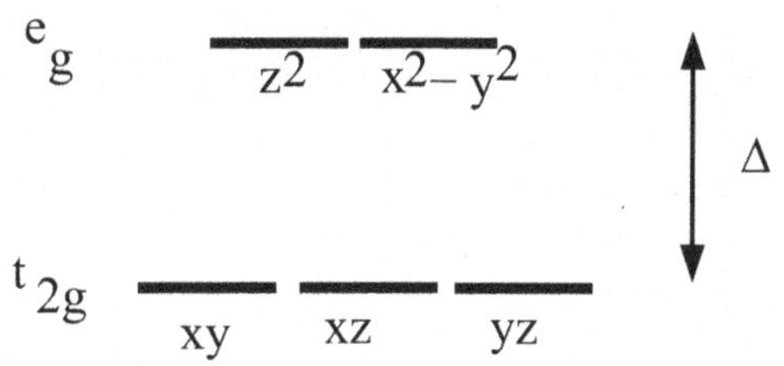

Fig. 8.5 Splitting of energy of d-orbitals in e_g manifold and t_{2g} manifold

8.4 World of Colors

We are surrounded by beautiful colors, really beautiful. Our clothes have beautiful colors on them. Plants are green and blood is red. What is the source of this color? Clothes have dyes containing pigments, like paints have pigments. These pigments are primarily inorganic in the sense they have a transition metal element in them. Transition metal elements are the one found in the center of the periodic table that have electrons in their d-orbitals. These include, for example, cobalt (Co), cadmium (Cd), chromium (Cr), manganese (Mn), etc. For example, cobalt (atomic number 27) has electronic configuration $1s^2 2s^2 2p^6 3s^2 3p^6 3d^7 4s^2$. The d-orbitals are fivefold degenerate. These orbitals are $d_{z^2}, d_{x^2-y^2}, d_{xy}, d_{yz}, d_{xz}$. However, in a transition metal compound, binding with other atoms called ligands, this degeneracy gets broken. We have orbitals $d_{z^2}, d_{x^2-y^2}$ called e_g manifold at higher energy than the orbitals d_{xy}, d_{yz}, d_{xz} called t_{2g} manifold, as shown in Fig. 8.5

The energy difference $\Delta = \hbar\omega_0$ is sub-eV and corresponds to visible wavelength. When we shine light, the right color is absorbed, rest scattered back, which we see. We see complimentary colors.

8.5 Scattering and Blue Sky

Why is the sky blue? From (8.16), blue light is more detuned and hence scatters less and reaches us (from sun above us), red light gets backscattered as less detuned. In the evening time, we mainly see scattered light (sun on the side), hence red light.

8.6 Problems

1. For dipole transition $d = 1\,\text{A}^\circ$ and $E = 1000\,\text{V/m}$, calculate the atomic absorption rate (rabi frequency) on resonance.
2. For dipole transition $d = 1\,\text{A}^\circ$ and $E = 1000\,\text{V/m}$, calculate the scattering rate (rabi frequency) for 1 GHz detuning.

3. Calculate the spontaneous emission rate for $d = 1$ A° and transition frequency $\omega_0 = 10^{15}$ Hz.
4. Calculate the Purcell spontaneous emission rate for $d = 1$ A° and $\omega_0 = 10^{15}$ Hz with $Q = 0.1$.
5. Calculate the optical susceptibility with $d = 1$ A°, $E = 1000$ V/m, 1 GHz detuning, and 1 ns decay rate.

References

1. A. Abragham, *Principles of Nuclear Magnetism* (Oxford University Press, Oxford, 2006)
2. C.A. Balanis, *Antenna Theory* (John Wiley and Sons (ASIA), Singapore, 2002)
3. L. Boithias, *Radio Wave Propagation* (North Oxford Academic Publishers, Oxford, 1987)
4. D.K. Cheng, *Field and Wave Electromagnetics* (Pearson Education, Delhi, 2002)
5. W.N. Christiansen, J.A. Hogbom, *Radio Telescopes (Monographs of Physics)* (Cambridge University Press, Cambridge, 1985)
6. R.E. Collin, *Antennas and Radio Wave Propagation* (McGraw Hill, New York, 1985)
7. A. Das, S.K. Das, *Microwave Engineering* (McGraw Hill, New Delhi, 2015)
8. R.S. Elliott, *Antenna Theory and Design* (Prentice-Hall, New Jersey, 1981)
9. R.P. Feynman, R.B. Leighton, M. Sands, *The Feynaman Lectures on Physics*, vol. II (Narosa Publishing House, New Delhi, 1986)
10. J. Griffiths, *Radio Wave Propagation and Antennas* (Prentice Hall International, London, 1987)
11. D. Griffiths, *Introduction to Electrodynamics* (Prentice Hall, Hoboken, 1999)
12. A.R. Harish, M. Sachidananda, *Antennas and Wave Propagation* (Oxford University Press, Oxford, 2007)
13. E. Hecht, *Optics* (Pearson, London, 2014)
14. P. Horowitz, W. Hill, *Art of Electronics* (Cambridge University Press, Cambridge, 1989)
15. E.C. Jordan, K.G. Balmain, *Electromagnetic Waves and Radiating Systems* (Prentice Hall of India, New Delhi, 2004)
16. J.D. Kraus, *Radio Astronomy* (McGraw Hill, New York, 1966)
17. J.D. Kraus, *Antennas* (McGraw Hill, Singapore, 1988)
18. D.M. Pozar, *Microwave and RF Wireless Systems* (John Wiley and Sons, New Jersey, 2001)
19. D.M. Pozar, *Microwave Engineering* (John Wiley and Sons, Singapore, 2003)
20. T. Pratt, C.W. Bostian, *Satellite Communication* (Wiley, Hoboken, 1986)
21. E.M. Purcell, D. Morin, *Electricity and Magnetism* (Cambridge University Press, Cambridge, 2013)
22. V. Ramachandran, K. Shankar, R.K. Shevgaonkar, *Transmission Lines and Networks* (Tata McGraw-Hill, New York, 1994)
23. S. Ramo, J.R. Whinnery, T. Van Duzer, *Fields and Waves in Communication Electronics* (John Wiley and Sons, Singapore, 2004)
24. M.O. Scully, M.S. Zubairy, *Quantum Optics* (Cambridge University Press, Cambridge, 1997)
25. R.K. Shevgaonkar, *Electromagnetic Waves* (Tata McGraw Hill, New Delhi, 2006)
26. M.I. Skolnik, *Introduction to Radar Systems* (McGraw Hill, New York, 1962)
27. A.W. Snyder, J.D. Love, *Optical Waveguide Theory* (Chapmann and Hall, London, 1983)

N. Khaneja, *Electromagnetic Waves and Optics*,
https://doi.org/10.1007/978-3-031-76144-7

28. W.L. Stutzman, G.T. Thiele, *Antenna Theory and Design* (John Wiley and Sons, Hoboken, 1987)
29. D. Suter, *The Physics of Laser-Atom Interaction* (Cambridge University Press, Cambridge, 1997)
30. S.M. Sze, *Semiconductor Devices: Physics and Technology* (Wiley, Hoboken, 1985)
31. D.R. Wehner, *High Resolution Radar* (Artech House, Boston, 1995)

The manufacturer's authorised representative in the EU is Springer Nature Customer Service Centre GmbH, Europaplatz 3, 69115 Heidelberg, Germany. If you have any concerns regarding our products, please contact ProductSafety@springernature.com

Printed and bound by CPI Group (UK) Ltd, Croydon, CR0 4YY
01/10/2026
02225977-0005